Comenuini

TECHNICAL CONTACTS

Other related titles of interest include:

BRIEGER, N. and J. Comfort
*Early Business Contacts**
BRIEGER, N. and J. Comfort
*Developing Business Contacts**
BRIEGER, N. and J. Comfort
*Advanced Business Contacts**
BRIEGER, N. and J. Comfort
*Social Contacts**
BRIEGER, N. and A. Cornish
*Secretarial Contacts**
BRIEGER, N. and J. Comfort
The Language of Business English
ADAMSON, D.
*Starting English for Business**
THORN, M. and A. Badrick
An Introduction to Technical English

* includes audio cassette(s)

TECHNICAL CONTACTS

Materials for developing listening and speaking skills for the student of Technical English

NICK BRIEGER JEREMY COMFORT

Prentice Hall

New York London Toronto Sydney Tokyo Singapore

PRENTICE HALL INTERNATIONAL ENGLISH LANGUAGE TEACHING

First published 1987 by
Prentice Hall International (UK) Ltd
Campus 400, Maylands Avenue
Hemel Hempstead
Hertfordshire HP2 7EZ
a division of
Simon & Schuster International Group

Printed and bound in Great Britain at the
University Press, Cambridge

Library of Congress Cataloging-in-Publication Data

Brieger, Nick.
Technical contacts.
Bibliography: p.
Includes index.
1. English language – Text-books for foreign speakers. 2. English language – Technical English.
3 Readers – Technology. I. Comfort, Jeremy.
II. Title
PE1128.B674 1987 428.2′4′0246 86-30374

British Library Cataloguing in Publication Data

Brieger, Nicholas
Technical contacts.
1. English language – Text-books for foreign speakers. 2. English language – Technical English.
I. Title II. Comfort, Jeremy
428.2′4 PE1128

ISBN 0-13-898263-5

6 7 8 9 10 97 96 95 94 93

Contents

The contents list below indicates the topic themes for each unit. Units 1–14 are functional/notional; Units 15–24 are grammatical.

Acknowledgements

The authors and publishers would like to thank the following for permission to reproduce copyright material:

The Sunday Times for an illustration from their issue of June 9, 1985;
Rohde & Schwarz GmbH & Co., Munich, West Germany;
Michael Joyce Consultants Limited;
British Telecommunications plc for the use of illustrations from *The Microchip Revolution*;
B.T. Batsford Ltd for an illustration from Jack Stroud Foster, *Structure and Fabric, Part 1*, The Mitchell Publishing Company;
The Yorkshire Evening Press;
Pitney Bowes plc for the illustrations on page 50;
Philips Electronics for illustrations from the instructions for the use of the Philips VR6460;
Mannesman Demag, Materials Handling Division;
The Financial Times.

Every effort has been made to trace and acknowledge ownership of copyright. The publishers will be glad to make suitable arrangements with any copyright holders whom it has not been possible to contact.

To Magda and Thérèse

Introduction

TARGETS AND OBJECTIVES

These materials are aimed at students who have a professional need for English in technical/industrial fields; people either in, or training for, posts as engineers or technicians. Some of the fields covered are: data processing, electronics, telecommunications, and mechanical engineering. For a full list, see the Contents page.

More specifically, this listening-based material is relevant for learners, at a *pre-intermediate level and above*, who need to revise or further practise the language around:

1. key functional/notional areas (Units 1–14)
2. key grammatical areas (Units 15–24)

The materials are designed to develop *listening skills* in the areas of:

1. extracting relevant information
2. structuring information
3. inferring meaning from context
4. becoming accustomed to different varieties (formal and informal) and different accents of English.

The materials also develop *speaking skills* around:

1. problem-solving activities
2. role-plays
3. discussion topics

ORGANIZATION OF MATERIAL

There are 24 units (see Contents page). Each unit consists of:

Introduction
A short written introduction to the topic of the unit, often introducing key vocabulary.

1 Listening
A taped listening passage, accompanied by an information transfer task.

2 Presentation
Highlighting and explanation of language items from the listening passage.

3 Controlled Practice
Exercises designed to practise the language items introduced in the Presentation.

4 Transfer
Pairwork, or occasionally group work, designed to encourage the students to use the language introduced and to be practised in a freer context.

5 Word Check
A glossary of the technical vocabulary that appears in the listening passage.

In the second part of the book is the Key Section for each unit. This contains:

Listening (1)
A tapescript and answers to the information transfer task.

Controlled Practice (3)
Answers to the controlled practice exercises.

Transfer (where necessary) **(4)**
Information for pair work activities.

The **Vocabulary Index** at the back of the book provides an alphabetical list of all words which appear in the unit glossaries (Word Check) and the unit numbers of their appearances.

THE ROLES OF THE TEACHER AND THE STUDENTS

The materials provide the teacher with an opportunity to strike a balance between two classroom roles: teacher-controlled and teacher-monitored.
They also give students an opportunity for autonomous learning (self-access).
Sections 1, 2, 3, and 5 (Listening, Presentation, Controlled Practice, and Word Check) can be done with or without a teacher.
Section 4 (Transfer) can be done by students in pairs or groups without a teacher, but some form of teacher monitoring is advisable.

Teacher's Notes

USES OF MATERIAL

1. As supplementary material to a General English course for students with an interest in or a need for technical English.
2. As extensive course material for the English component in a Technical Training Course.
3. As a self-study/homework component for a Technical English course.
4. As follow-up material on completion of a Technical English course.

SELECTION OF MATERIAL

The units are *not* graded. Therefore, teachers should select according to:

1. Topic (see Contents page)
2. Language Area (see Contents page)

USING A UNIT

Introduction

You can use the text as a basis for presenting the unit orally or for asking students to read it through themselves. The words in *italic type* are key vocabulary items to which they will be exposed in the listening passage. Therefore, it is important that they understand the meaning and recognize the pronunciation.

1 Listening

- **i** Prepare the students for the listening task. Make sure they are absolutely sure of what they have to do.
- **ii** Play the tape right through, without stopping.
- **iii** For many students it will be necessary to play the tape again, stopping it at appropriate places.
- **iv** Let the students check their answers with the Key.
- **v** Play the tape a third time if there are major differences between the Key and student answers.
- **vi** Refer the students to the Word Check (Section 5) if there are vocabulary problems.

2 Presentation

- **i** Ask the students to read through the presentation and explanation of the language area.
- **ii** Get them to give you additional examples of the language presented.
- **iii** If necessary, look at the tapescript in the Key to identify exponents of the language presented.

3 Controlled Practice

- **i** Ask students to complete the exercises and then check their answers with the Key.
- **ii** Advise on alternative answers or give more practice where necessary.

4 Transfer

Most of these exercises involve pairwork.

i Divide the class into pairs.

ii Assign roles (Student A and Student B). Make sure they only look at their role/information (Student B's information is always in the Key Section).

iii Monitor the pairs while they carry out the transfer activity, prompting the use of practised language if necessary.

5 Word Check

This can be used during the listening activity, if necessary.

The glossary only provides definitions, since examples would necessarily be too subject specific. Students should be encouraged to provide their own examples from their own fields.

NOTE

The following symbols have been used to indicate what is missing in the exercises:

– – – – – – one or more words;

__________ only one word.

Notes to the Student

WHO IS IT FOR?

This material is for students who have some previous knowledge of English and wish to practise it in a technical context. It can be used by students working alone; as self-study or homework material during a Technical English course; or as follow-up material after a Technical English course.

SELECTION OF MATERIAL

You can work through the material starting at Unit 1. Alternatively, you can choose units on the basis of the topic or language area (see Contents page). There are two main language areas:

1 *Functional*: Units 1 to 14. These give practice in the *use* of the language to describe, explain, demonstrate, etc.
2 *Grammatical*: Units 15 to 24. These give practice in important areas such as the use of tenses, prepositions, formation of questions, etc.

USING A UNIT

All of the units except the Transfer Section can be done without a teacher.

Introduction

This tells you something about the subject of the unit. If necessary, it introduces some important vocabulary which you will find in *italics*.

1 Listening

All the listening activities have an exercise with them. These exercises will help you to listen more actively. Follow this procedure:

i Read through the introduction to the listening. Make sure you understand what you must do as you listen.
ii Listen to the tape without stopping it.
iii As you listen, try to do the exercise.
iv If necessary, listen again. This time stop the tape and replay sections if you need to.
v Check your answers with the Key at the back of the book.
vi If your answers are wrong, listen again or check the tapescript in the Key. Use the Word Check if you cannot understand some of the words.

2 Presentation

i Read carefully the presentation and explanation of the language areas.
ii Try to remember how this language was used on the tape. If you wish, play the tape again.

3 Controlled Practice

i Complete the exercises.
ii Check your answers with the Key.

iii If your answers are wrong, look again at the Presentation and try to see why you have made mistakes.

4 Transfer

This is best done with the teacher. But you can do the pairwork with a colleague. Follow this procedure:

i Decide who is Student A and who is Student B.
ii Student A should *only* look at the Student A copy.
iii Student B should *only* look at the Student B copy in the Key Section.
iv Carry out the Transfer activity. Try to use the language you have learnt.

5 Word Check

The words are taken from the listening passage. Definitions are given. Try to think of examples of how you could use these words in your own field.

NOTE

We have used the following symbols. They show you what is missing in the exercises:

– – – – – – one or more words;
__________ only one word.

UNIT 1 A Bridge or a Tunnel?

(dimensions and specifications)

Introduction

This unit gives details of 5 plans to build a cross-channel link either by a bridge, a tunnel or a bridge/tunnel. The presenter outlines the *specifications* for each *proposed structure* and, in particular for a bridge, considers the number of *spans* needed.

1 Listening

Listen to the tape which gives part of a presentation on the 'Brunnel Project'. The project is to build a road link across 30 kilometres of sea. As you listen, number the plans below according to the number given by the presenter.

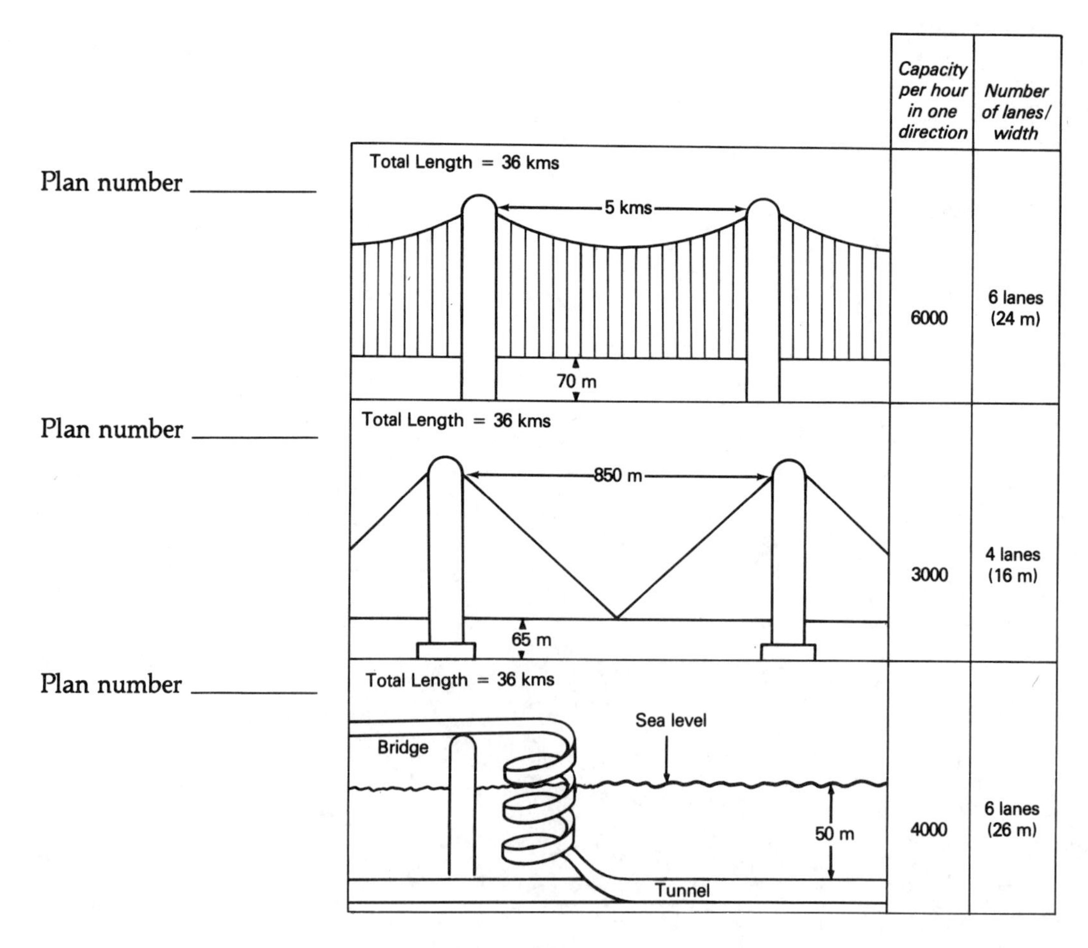

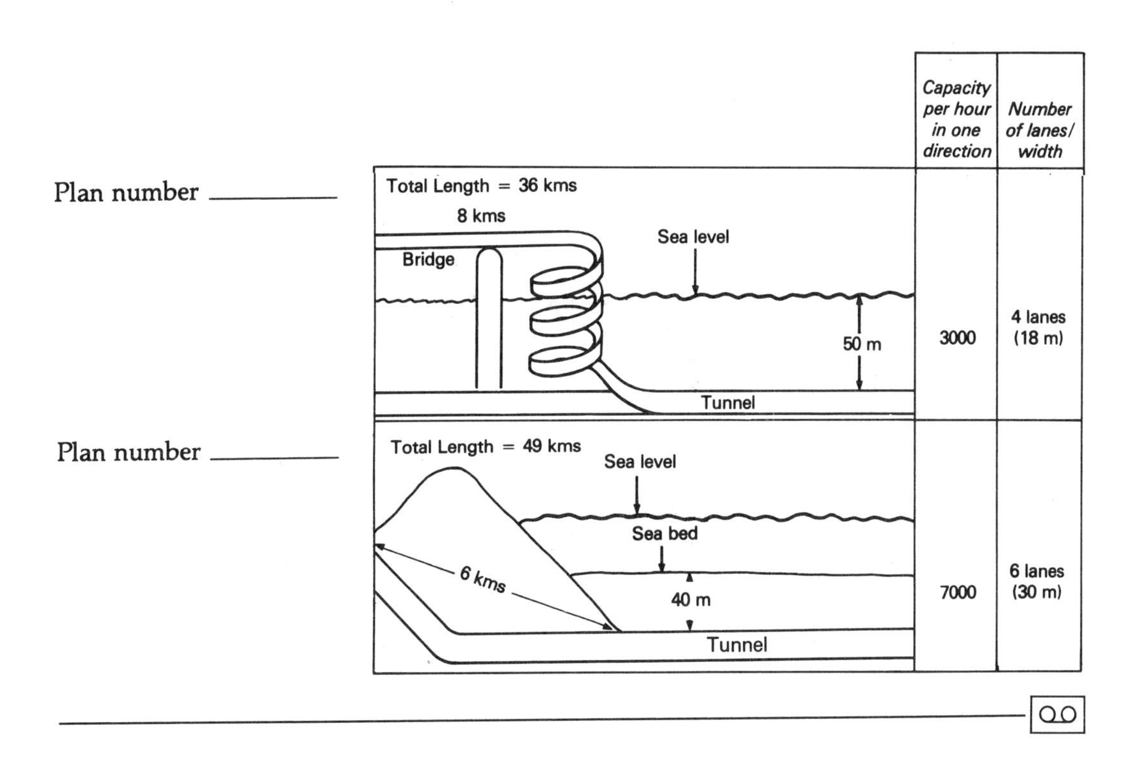

2 Presentation

Now look at the diagram below. Notice how the expressions are used to describe the specifications and capacity of the structure.

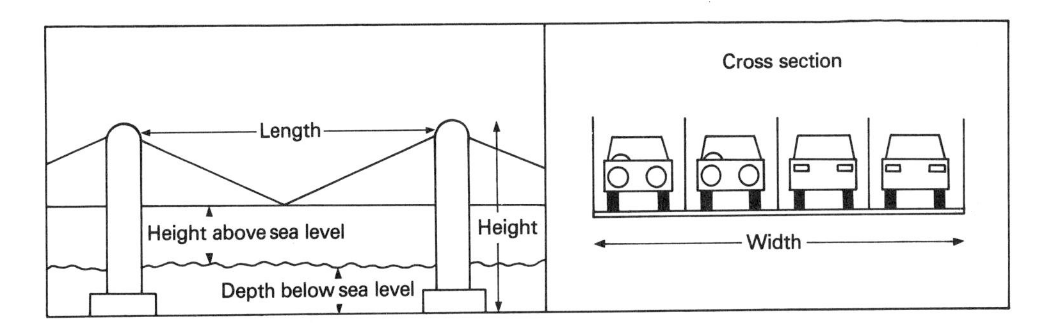

Specifications

850 metres *long*	850 metres *in length*	*the length* of each span is 850 metres
18 metres *wide*	18 metres *in width*	*the width* of the motorway is 18 metres
200 metres *high*	200 metres *in height*	*the height* of each pillar is 200 metres

Location

The motorway is *at a height of* 65 metres above sea level
The bottom of each pillar is *at a depth of* 150 metres below sea level
30 metres below the sea bed

Capacity
it *can/will carry* 300 cars per hour
carries

Notice the question forms:
How long is each span?
How wide is the motorway?
How high is each pillar?
How many cars can/will the motorway carry?
How much traffic does the motorway carry?

3 Controlled Practice

Look at the plans below. One is for a pedestrian subway (or underpass), and the other is for a pedestrian bridge (or overpass). Complete the sentences using the specifications and appropriate expressions from the Presentation section.

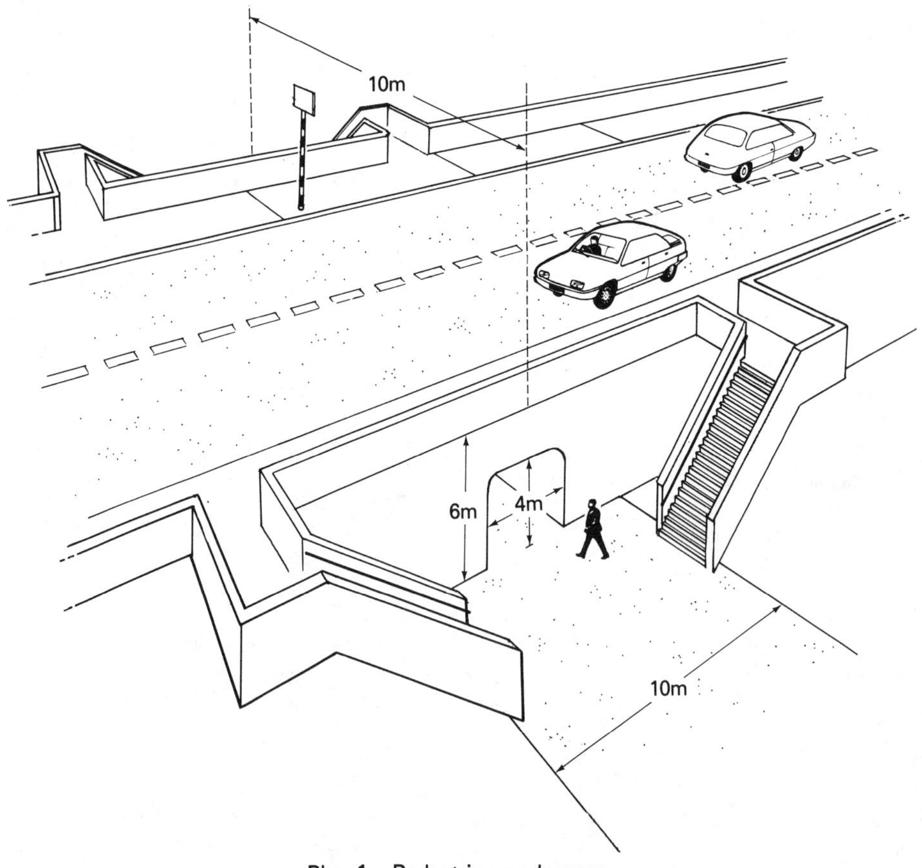

Plan 1 Pedestrian underpass

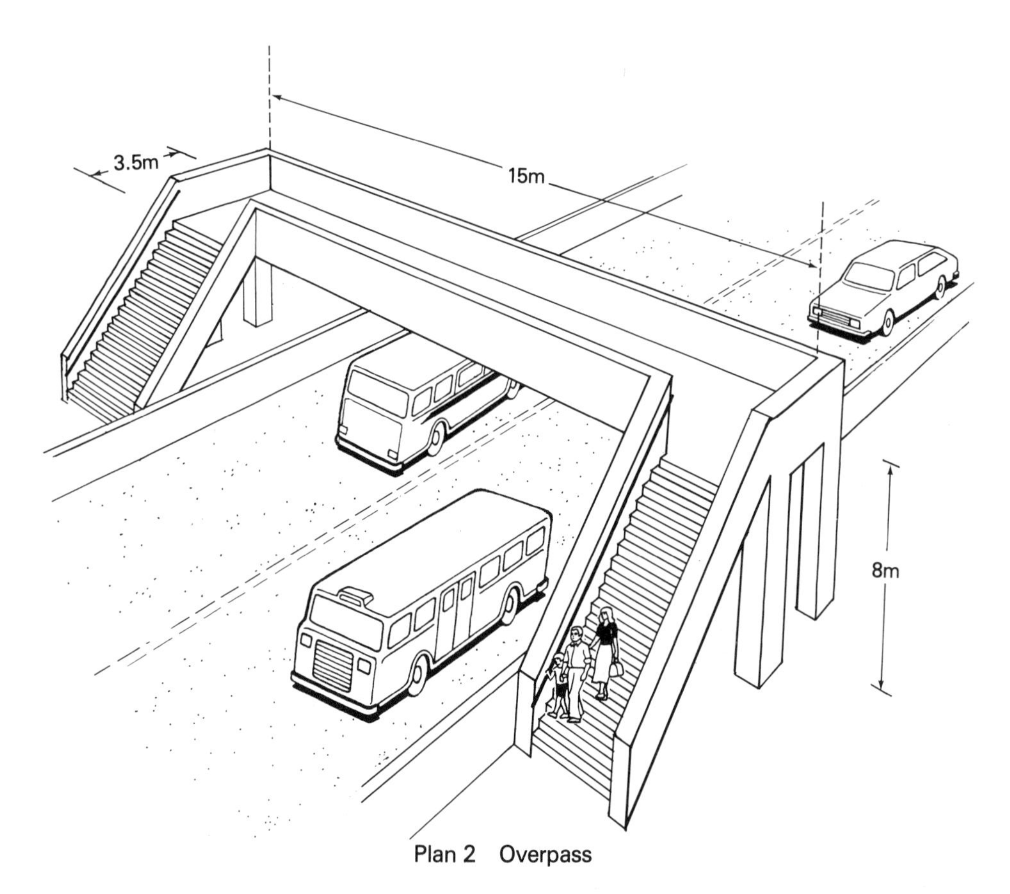

Plan 2 Overpass

Plan number 1 is for a pedestrian underpass, __ __1__ __ 6 metres below street level. The structure will consist of 25 steps on each side for access, and a tunnel, which will be __ __2__ __ in __ __3__ __ with overhead lighting. To provide enough space for the estimated maximum capacity, the __ __4__ __ of the tunnel will be 4 metres, and it will also be __ __5__ __.

Plan number 2 is for an overpass __ __6__ __/__ __7__ __ above street level. The construction will consist of a bridge supported by four pillars. Access will be via steps on each side. The walkway will be 15 metres __ __8__ __ and __ __9__ __ metres in __ __10__ __. As it will not be covered, the __ __11__ __ will, of course, be unlimited. The construction company estimates that the whole structure __ __12__ __ 300 pedestrians at one time.

4 Transfer

PAIRWORK

Student B: Turn to the Key Section.

Student A: You are a civil engineer and have designed a plan for a single-storey office block. Describe your plan and its specifications to your partner according to the diagram below. Your partner will draw your plan on his/her own graph. After your partner has completed the plan, compare diagrams.

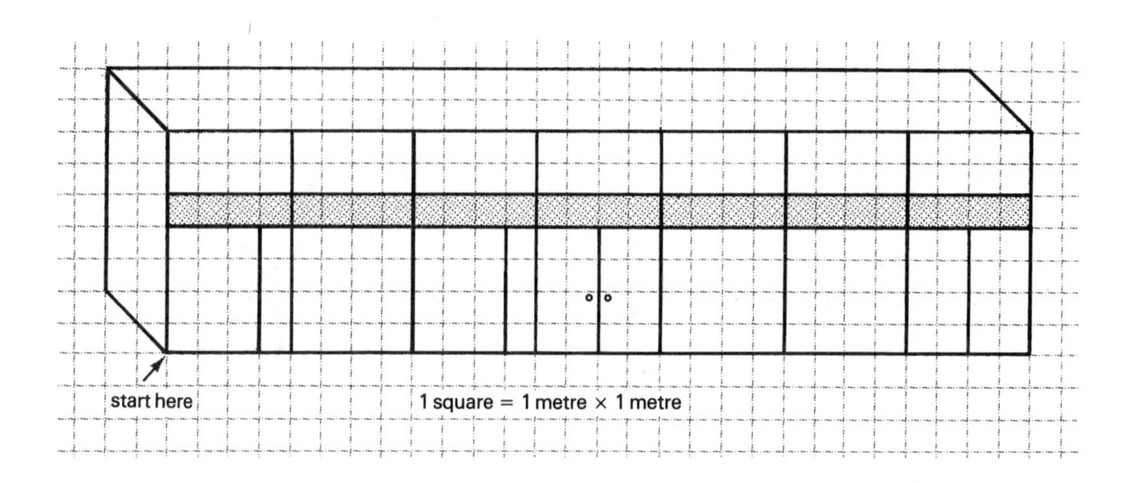

Student A: Now listen to your partner's plan for a two-storey office block. Draw the plan below according to your partner's specifications. After you have completed your diagram, compare it with your partner's version.

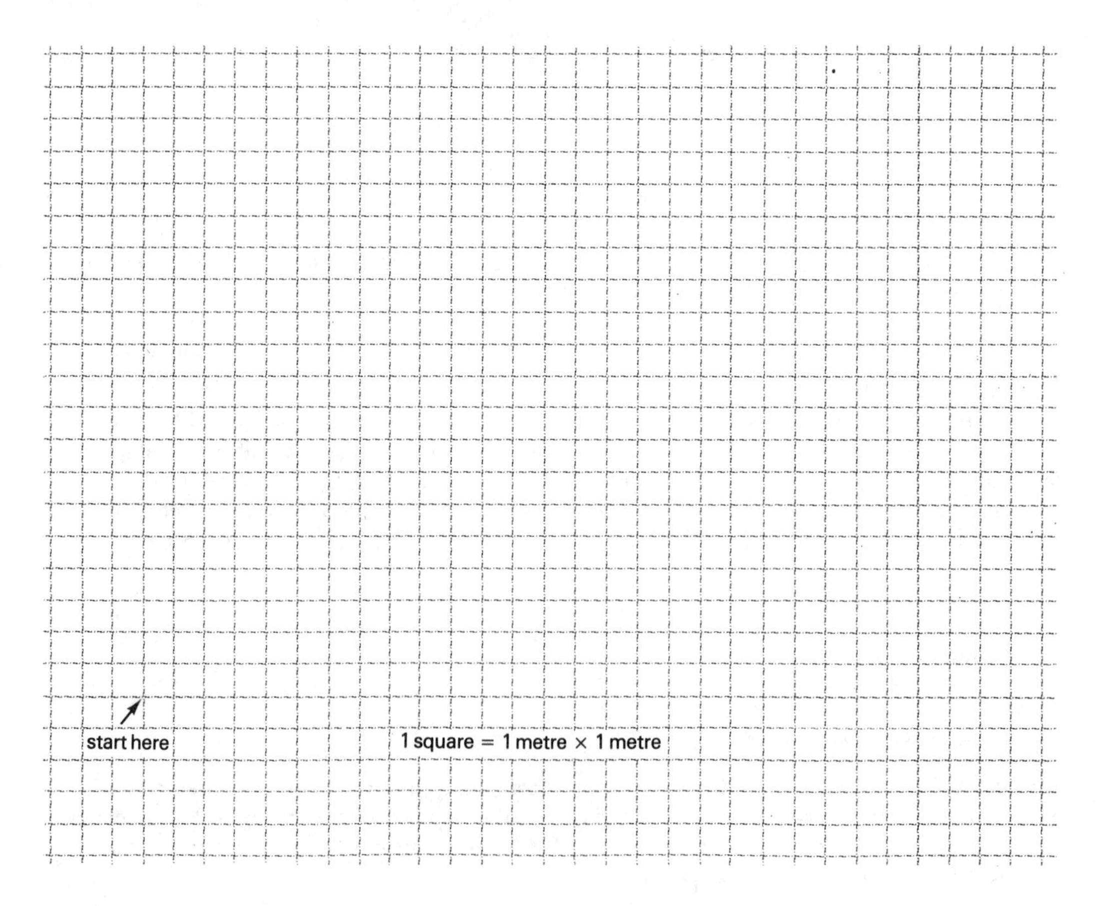

5 Word Check

ROADS AND TRANSPORT
passenger — person (other than the driver or crew) who travels by car, bus, boat, plane, etc
vehicle — means of transport, e.g. car, bus, lorry, etc.
goods vehicle — vehicle, e.g. lorry, that carries goods (heavy articles) as opposed to people
lane — any one of the parallel parts into which wide roads are divided
capacity — maximum amount or quantity
motorway — wide road for fast vehicles
to overtake — to pass a vehicle

PLANS
project — a plan for work
proposal — suggested plan

UNIT 2 Protecting your Computer System

(obligation and permission – modals)

Introduction

This unit looks at different ways of protecting a computer system. It considers how *access* can be limited to *authorized* users, and suggests ways of achieving this.

1 Listening

You are going to hear a discussion between a data processing manager and a computer consultant. The data processing manager is telling the computer consultant what security requirements his company needs for their computer equipment. The consultant has a checklist which shows different features suitable for a variety of systems. As you listen tick the appropriate column in the checklist. You need to decide whether the security feature is *Necessary, Unnecessary, Permitted/Possible* or *Not Permitted/Impossible* for the manager's computer system.

SECURITY CHECKLIST *Specifications*	*Necessary*	*Unnecessary*	*Permitted/ Possible*	*Not permitted/ Impossible*
only allow authorized access to system				
provide physical isolation for hardware				
protect both hardware and software				
indicate unauthorized access to system				
protect hardware in an open environment				
all company personnel have access				
all members of staff use all computer facilities				
control access via identification codes				
control level of access via passwords				
allow managers to enter, view and amend data				
allow operators to amend data				
only authorized personnel delete files				

2 Presentation

The table below shows the words used to express what is:

i. NECESSARY/NOT NECESSARY
ii. POSSIBLE/NOT POSSIBLE
iii. PERMITTED/NOT PERMITTED

NECESSARY	NOT NECESSARY
must	needn't don't } need to doesn't } need to

POSSIBLE	NOT POSSIBLE
can	mustn't can't

PERMITTED	NOT PERMITTED
may can	may not can't/cannot mustn't

Notes
need is used both as a modal and a full-verb.
As a *modal* it is normally used only in the negative and is followed by an infinitive without *to*, e.g. *We needn't provide total protection for hardware and software.*
As a *full-verb* it is followed by an infinitive with *to*, e.g. *But I need to know more about control of access.*

3 Controlled Practice

Now use the language from the Presentation, and the information from the table below, to make sentences about who can do what with the Company's computer files. You will find the correct verb by looking at the *Feature* section. The first one has been done for you.

File/System	*Operation*	*User*	*Feature*
Name & Address File	amend	keyboard operator	permitted
Name & Address File	update	senior manager	not necessary
Name & Address File	delete	keyboard operator	not permitted
Name & Address File	call up	any authorized user	permitted
Budget Forecast	amend	keyboard operator	not permitted
Budget Forecast	delete	company executive	not permitted
System	switch off	computer operators (in the evening)	not necessary
All File Disks	store away	computer operator (in the evening)	necessary
System	power up	computer operator (in the morning)	necessary
Supplier File	check	keyboard operator	not necessary
Supplier File	check	Purchasing Manager (before placing an order)	necessary
Debtor File	amend	keyboard operator (without the approval of the Accounts Manager)	not permitted

1. A keyboard operator may/can amend the Name & Address File.
2. _______________________________________
3. _______________________________________
4. _______________________________________
5. _______________________________________

6. ______________________________

7. ______________________________

8. ______________________________

9. ______________________________

10. ______________________________

11. ______________________________

12. ______________________________

4 Transfer

Discuss the priority you would give the following measures in ensuring the security of your computer installation.
a. stricter vetting of personnel in contact with computer equipment
b. stricter control of access to hardware
c. stricter control of access to software.

PAIR or GROUP WORK
Imagine that you are a data processing manager. Discuss how you would improve the security of your computer equipment.
Consider how you would control:

a. the personnel with access to the equipment
b. access to the hardware
c. access to the software

5 Word Check

EQUIPMENT
processing equipment — the central part of a computer system which performs the operations
terminal — a piece of equipment for giving instructions to or getting information from a computer
hardware — equipment in a computer system
software — programs for a computer system

PERSONNEL
keyboard operators — people employed to put information into, and extract information from, terminals

VERBS
to house — to locate
to input — to put information into a computer
to get into (the system) — to use the computer system illegally
to specify — to say exactly
to gain (access) — to get into the computer system legally
to enter (data) — to put in
to amend (data) — to change
to delete — to remove

UNIT 3 Designing a Computer System

(classification)

Introduction

This unit deals with the components of a computer system. It describes the different parts of a system and the *devices* used for *input*, *storage*, and *output*.

1 Listening

Paul Bailey has a small trading company. He would like to computerize the company's accounts, payroll, sales, purchasing and stock control. As he doesn't know very much about computers, he invites a computer consultant to advise him. As you listen complete the chart below.

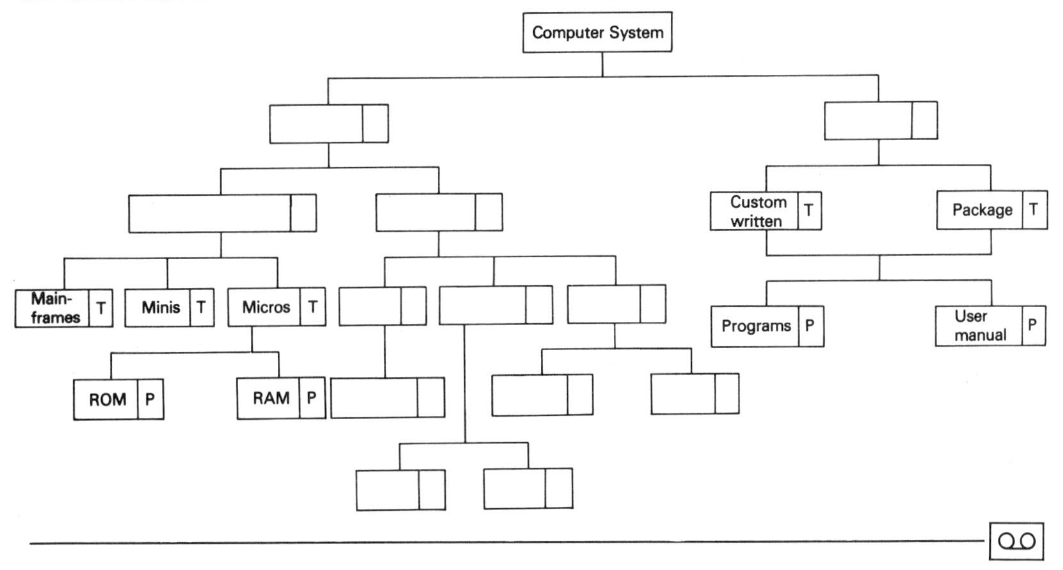

2 Presentation

a.i One way to identify objects is to classify them according to type, e.g. Mainframes, minis and micros are *types of* computers.
We can also use other expressions, as follows:
Mainframes, minis and micros are *kinds of* computers.

If we want to focus on only one of the items, we use the singular,
e.g. A mainframe is *a type of* computer.
A VDU is *a kind of* TV screen.

a.ii We can also use a number of verbs to express the same relationship.
e.g. We can *divide* computers *into* mainframes, minis and micros.
We can *split* computers *into* mainframes, minis and micros.
We can *classify* computers *as* mainframes, minis and micros.
Computers *fall into 3 types*: mainframes, minis and micros.

b.i Another way to identify an item is to describe its most important parts,
e.g. The 2 *parts of* a computer system are hardware and software.
If we want to focus on one of the items, we can use the singular,
e.g. Hardware is *one component of* a computer system.

b.ii We can also use a number of verbs to express the same relationship,
e.g. A computer system { *consists of* / *comprises* } hardware and software.

3 Controlled Practice

A. Now look back at your chart in the Listening Section. Where there is a box, indicate T for type and P for part.

B. Now complete the sentences below using the information from your chart. Put one word in each space.

1. A computer's hardware ________ a CPU and peripheral equipment.
2. Input, output and storage devices are ________ ________ peripheral equipment. Storage devices ________ ________ 2 categories — tape or disk.
3. We can ________ a printer and a VDU ________ output devices.
4. We ________ ________ software ________ 2 types — custom written and package.

4 Transfer

PAIRWORK

Student B: Turn to the Key Section.

Student A: Below you will find a chart which describes the ABC Model A computer. Describe it to your partner, who will draw a chart according to the information you give. After your partner has completed the chart, compare your versions.

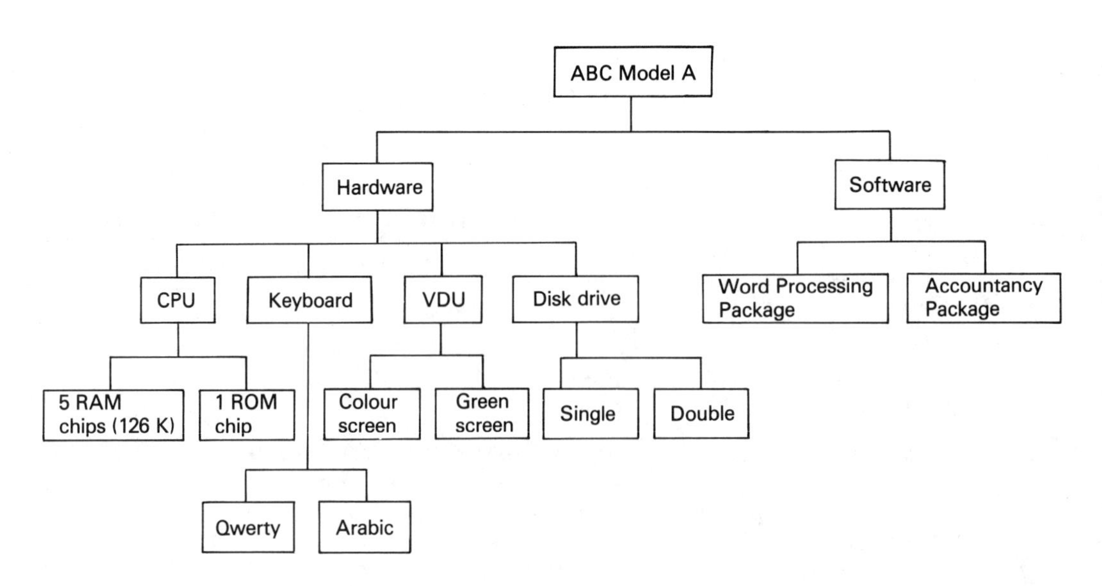

Student A: Now listen to your partner's description of the ABC model B. Complete the chart below according to the information you hear. After you have completed your chart, compare it with your partner's version.

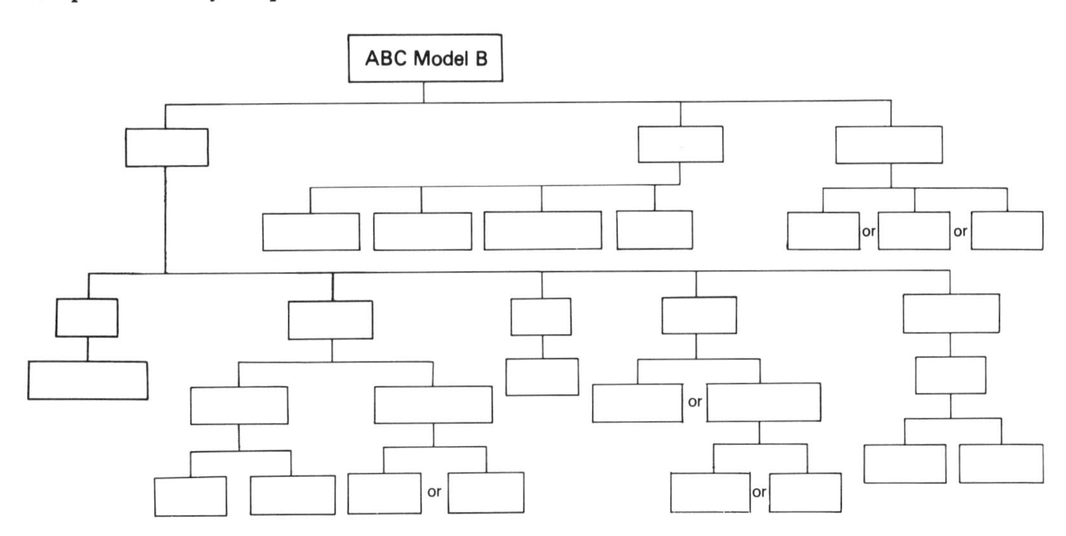

5 Word Check

EQUIPMENT
hardware — equipment in a computer system
software — programs for a computer system
peripheral equipment — other parts of a computer system (not the CPU), e.g. printer, disk drive, etc.
central processing unit (CPU) — main part of a computer system
visual display unit (VDU) — computer input/output device with a screen and a keyboard
screen — front of a VDU (or TV) on which you can see information
printer — device which prints information
storage device — equipment on which to record information, e.g. disk drive
keyboard — device, like a typewriter, to input information into a computer

APPLICATIONS
accounts — details of money received and paid out
payroll — total amount of wages to be paid out to employees
wages — payment for work
purchasing — buying
stock control — check on quantity of goods in shop or warehouse

OPERATIONS
to process (information) — to put information into and get information out of a computer
to calculate — to find out something by using numbers
to load — to put information or a program into a computer (from disk or tape)
to multiply — mathematical calculation (symbol ×)
to add — mathematical calculation (symbol +)
to subtract — mathematical calculation (symbol −)
to store — to save information for later use, e.g. on disk or tape

UNIT 4 Optical Fibres

(quantity and amount)

Introduction

This unit deals with a new material used for transmitting sound and data: optical fibre. Optical fibre is made of glass and uses light (usually from a laser) to transmit messages.

1 Listening

An engineer is talking to a journalist about the advantages of optical fibre compared with conventional copper cable.
As you listen, match the advantage with its effect(s). The first one has been done for you.

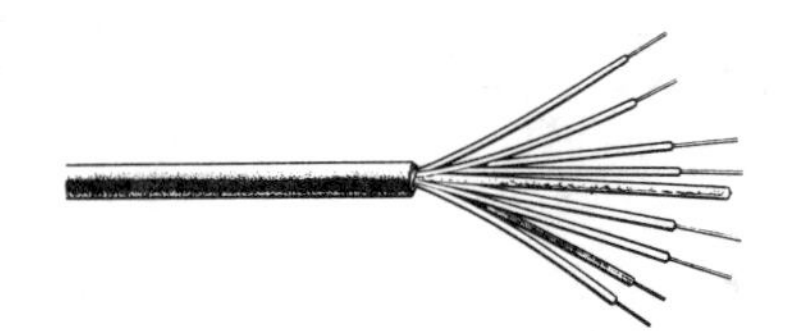

ADVANTAGE	EFFECT
1. Higher capacity	a. less frequent repeaters
2. Lower material cost	b. more security
3. Smaller size	c. cheaper to produce
4. Higher quality of transmission	d. more information
5. Complete electrical isolation	e. less space in ducts
	f. less interference/crosstalk

2 Presentation

In his talk, the engineer compared QUANTITY and AMOUNT:

QUANTITY
With conventional cable, you need *many more ducts*
With conventional cable, you can transmit *far fewer telephone calls*

AMOUNT
Optical fibres can carry *much more information*
Optical fibres take up *much less space*

Summary
With nouns you can count (QUANTITY) e.g. tables, cables, etc.

Strengthener	*Adjective*	*Noun*
Many	more	cables
Far	fewer	repeaters

With nouns you cannot count (AMOUNT) e.g. time, space, etc.

Strengthener	*Adjective*	*Noun*
Much	more	space
Far	less	time

3 Controlled Practice

A. Classify the following nouns as either COUNTABLE (C) or UNCOUNTABLE (U)

1. telephone call
2. repeater
3. information
4. data
5. duct
6. money
7. security
8. interference
9. crosstalk
10. space
11. capacity
12. equipment

B. Complete these sentences

1. Optical fibres carry ________ ________ information than conventional cables.
2. ________ ________ telephone calls can be transmitted using optical fibre.
3. ________ ________ data can be transmitted using conventional cable.
4. You hear ________ ________ crosstalk when using optical fibre.
5. There is ________ ________ interference on copper cables.
6. Optical fibres occupy ________ ________ space.
7. You need ________ ________ ducts or cable pipes with optical fibre.
8. Conventional cable has ________ ________ capacity than optical fibre.
9. Conventional cable will cost ________ ________ to produce in the future than optical fibre.
10. You need ________ ________ equipment, such as repeaters, on a copper cable line.

4 Transfer

Draw up a list of the advantages of electronic components over electromechanical components (e.g. the microchip versus the relay). Then present and explain the advantages.

5 Word Check

capacity — how much something can hold e.g. the capacity of the truck is 200 cases.
material cost — cost of the raw material (e.g. glass)
ducts/pipes — channels for carrying cables
to amplify — to increase the strength of the signal
signal loss — decrease in the strength of a signal
repeater/booster — type of amplifier
interference — a disturbance to the signal caused by unwanted signals
crosstalk — a type of interference — sounds of another telephone call on the line
electrical isolation — outside electrical signals can*not* interfere with the signal
security — the information/data cannot be changed, accessed by other users
to corrupt (data) — to change or delete data

UNIT 5 Shape Memory Alloys

(cause and effect)

Introduction

This unit deals with applications of a new metal *alloy* (a combination of two elements). As the name suggests, this alloy can remember its *shape* (form) and will always return to it after heating.

1 Listening

A research engineer is talking about the applications of a new type of metal alloy. The table below shows three applications and should show the sequence of actions and reactions. As you listen, compete the table.

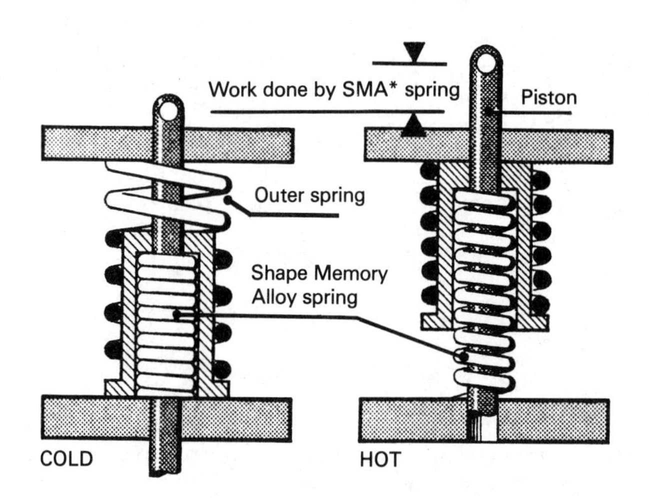

Application	*Cause*	*Primary effect*	*Secondary effect*	*Tertiary effect*
1. Piston	Piston cold	_ _ _ _	piston doesn't move	
	Piston _ _ _	Coil expands	_ _ _ _	
2. Coffee Machine	_ _ _ _ _ _ _ _	SMA actuator opens valve	_ _ _ _ _ _ _ _	
3. Cooling fan in car	Cold weather	_ _ _ _	_ _ _ _	car warms up
	_ _ _ _	_ _ _ _	_ _ _ _	_ _ _ _

* SMA is an abbreviation for Shape Memory Alloy

2 Presentation

During his talk, the Research Engineer expressed the concepts of CAUSE and EFFECT in three different ways:

a. *When* clause
   e.g. *When* the piston *is* cold, the coil *contracts*.

b. *Verbs* linking the cause to the effect
   e.g. Heat *causes* it *to* expand. (cause something *to* happen)
   This *makes* the fan close. (*make* something happen)
   Cold weather *leads to* expansion. (lead to something)
   results in (result in something)
   brings about (bring about something)

c. *Adverbs* linking the cause to the effect
   e.g. The spring contracts and *so* allows the fan to operate.
   therefore
   consequently
   thus
   as a result

3 Controlled Practice

A. Use the above three ways of expressing cause and effect, and the table from Section 1, to finish these sentences.

1. ________ the coffee machine ________ ________ ________ ________, the SMA actuator ________ the valve and ______ the water ________ ________ ________ ________ ________.
2. ________ it ________ ________, the spring ________. This ________ the fan ________ ________ and ________ the car ________ ________.
3. In warm weather, the spring ________ and ________ the fan ________. This ________ the car ________ ________ ________.

B. Use the diagram of a process cycle to complete the description:

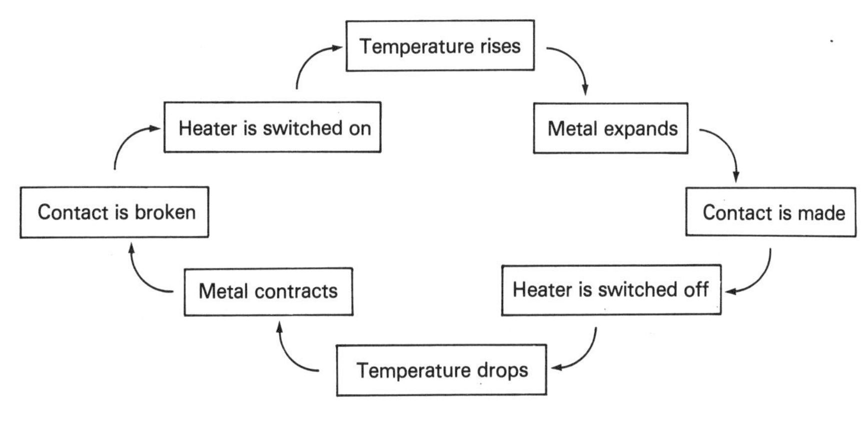

Description of process cycle
____ the temperature ____, ____ ____ and ____
____ ____. This ____ the heater to be switched off and
____ ____ ____. In time, the metal contracts and ____
____ broken. The broken contact ____ the heater ____
____ ____ ____ and ____ the cycle starts again.

4 Transfer

Think of other applications of Shape Memory Alloys and explain how they would work.
E.g. Air conditioning systems
An electric toaster
An electric iron
etc.

5 Word Check

PROCESS VERBS
to deform – to change shape
to contract – to get smaller
to expand – to get bigger
to supply – to give, provide
to give off – to transmit into the atmosphere
to warm (up) – to become warm, cf. to heat
to cool (down) – to become cold
to dump – to drop, to empty the contents
to fit – to attach, to fix

DEVICES
a valve – a control which opens and closes
an actuator – a control which activates (something)
a fan – a device which blows out hot or cold air

UNIT 6 What is a Transistor?

(properties)

Introduction

This unit deals with two aspects of a transistor:

a. the basic raw material – silicon. It explains the properties of this element.
b. the composition of a transistor – it explains how two types of silicon are used to create a transistor:
1. p.type to which boron is added,
2. n-type to which phosphorus is added.

1 Listening

Listen to the electronics engineer talking about transistors. As you listen, complete Figures 1, 2 and 3.

a. ____	49%
b. ____	26%
Aluminium	8%
Iron	4%
Calcium	3%
Sodium	2%
Potassium	2%
Magnesium	2%
Hydrogen	1%
All others	3%

Fig. 1 Principal Elements

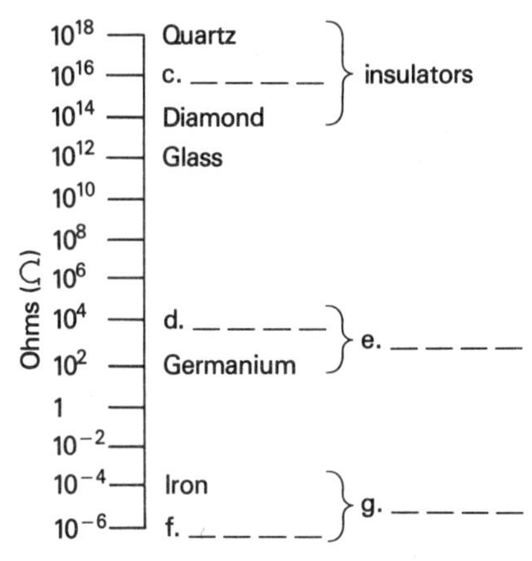

Fig. 2 Resistance of materials

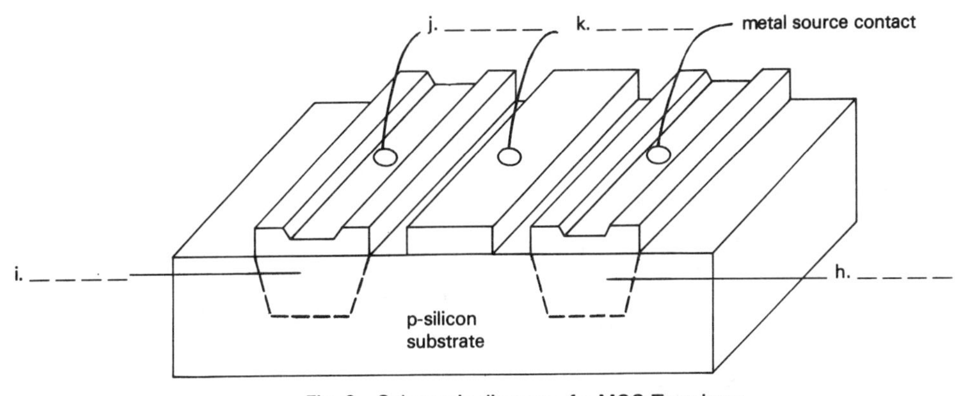

Fig. 3 Schematic diagram of a MOS Transistor

2 Presentation

The electronics engineer described the following:

What is it?	It's *a* solid/liquid/gas
What can it do?	It *can* increase conductivity It *has the ability to* conduct electricity It *allows/doesn't allow* electricity to flow
What is it used for?	It's *used for* insulat*ing*
What is it called?	They'*re called* semiconductors It's *known as* n-type silicon It's *termed* p-type silicon
What is it composed of?	It *is composed of* silicon It *consists of* two parts It *is made of* germanium A p-n junction *is formed*
Where is it?	At the top On the left — In the middle — On the right At the bottom

3 Controlled Practice

Complete the sentences using the language presented above.

A.
1. Silicon circuits __ __ __ __ __ __ __ __ __ __ __ __ __ transistors, diodes, resistors and capacitors.
2. The capacitor __ __ __ __ __ __ __ __ __ __ __ __ __ storing the electrical signal.
3. The capacitor __ __ __ __ __ __ __ __ __ __ __ __ __ two conductors separated by an insulator.
4. Capacitors, which __ __ __ __ sometimes __ __ __ __ condensers, can store electrical signal.
5. Diodes __ __ __ __ electricity to flow in one direction. They __ __ __ __ __ __ __ __ electricity to flow in the other.
6. __ __ __ __ __ __ __ __ __ __ __ __ __ of Figure 1 is the world's commonest element — Oxygen.
7. __ __ __ __ __ __ __ __ __ __ __ __ __ of Figure 2 are the semiconductors.
8. __ __ __ __ __ __ __ __ __ __ __ __ __, you can see Figure 1.
9. __ __ __ __ __ __ __ __ __ __ __ __ __, you can see Figure 2.
10. __ __ __ __ __ __ __ __ __ __ __ __ __, you can see Figure 3.

B. Complete the following description of COMPUTER AIDED DESIGN.

Computer Aided Design, often simply __ __1__ __ CAD, __ __ __ __2__ __ __ __ __ __ __ __ designing complex electronic chips. The system __ __ __ 3 __ __ __ __ hardware (a computer terminal, disk drives and a plotter) and software computer programs that can generate designs. The computer __ __ __ __ 4__ __ __ __ __ __ __ __ to produce 3-dimensional views of the circuit and then __ __5__ __ store the diagram. It __ __6__ __ the designer to call up the diagram at any time and make changes to it.

4 Transfer

PAIRWORK – GUESS WHAT IT IS

Student B: Turn to Key Section.

Student A: Think of an object (e.g. desk, telephone, light bulb, TV, etc.) and ask Student B to identify it by asking a series of questions, such as:

What is it?
What can it do?
What is it used for?
What is it composed of?

Student B now thinks of an object and the roles are reversed.
Note: Do *not* ask what it is called. You must decide from the description.

5 Word Check

PROPERTIES
solid – a substance that is hard, cf. liquid and gas
conductivity – the ability to let electric current flow
  to conduct
  a conductor
resistance – the ability to stop electric current from flowing (measured in ohms)
insulation – material used to resist the flow of electricity
  to insulate
  an insulator
an impurity – an additional substance/element found in a pure element
to dope – to add impurities
a dopant – an impurity

COMPOSITION
a substrate – a base material on which parts can be mounted/fixed
contact leads/electrodes – a connection through which electricity can flow
- metal source contact: electric current is introduced into the transistor through this lead
- metal drain contact: electric current is taken out through this lead
- gate electrode: electric current is controlled through this lead.

UNIT 7 Testing Circuits

(demonstrating)

Introduction

In this unit, a logic analyser is demonstrated. This device is used to test electronic circuits. A *probe* (a test connector) is connected to a circuit and then electrical data is shown on a screen. The operator can take a *sample* (a sort of photograph) of the transmission of data in the circuit and then analyse it.

During the demonstration, the *basic menu* (this shows how the analyser is set up) is displayed on screen. Then, a *time display* is shown (this displays 8 of the channels in a microprocessor circuit).

1 Listening

You are going to hear a demonstration of a Logic Analyser called IMAT. This device is used to test and analyse microprocessor circuits. As you listen, complete the labelling of the front panel of the IMAT. The demonstration is split into two parts:

1. Basic menu on screen
2. Time display on screen

1. Basic Menu

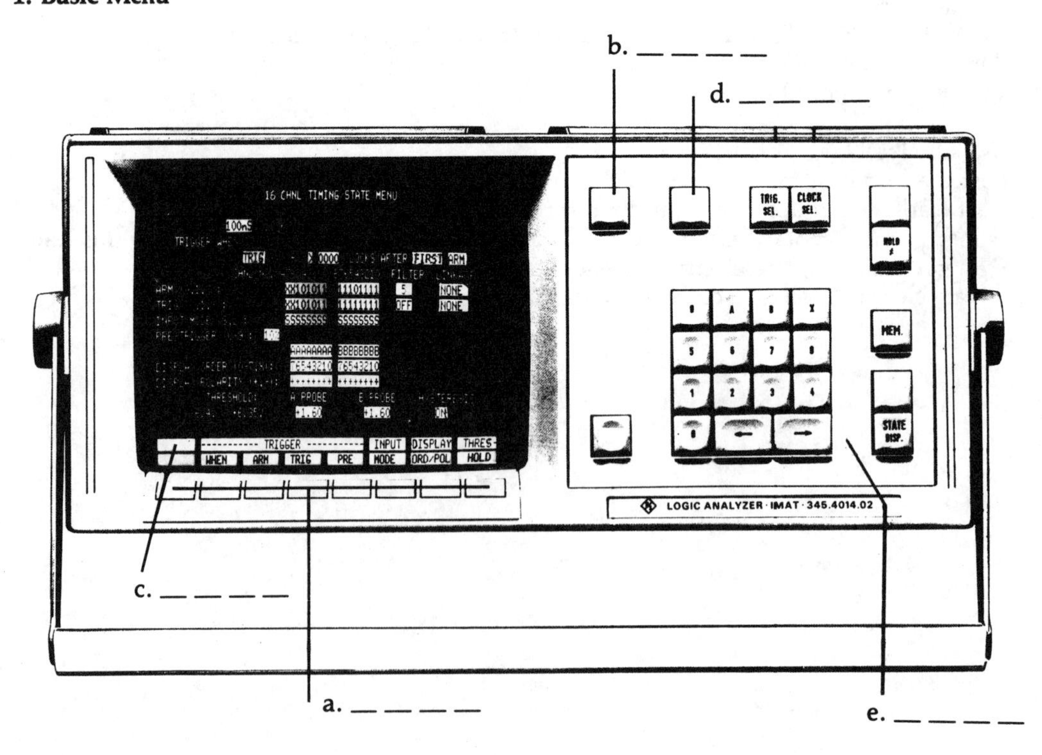

2. Time Display

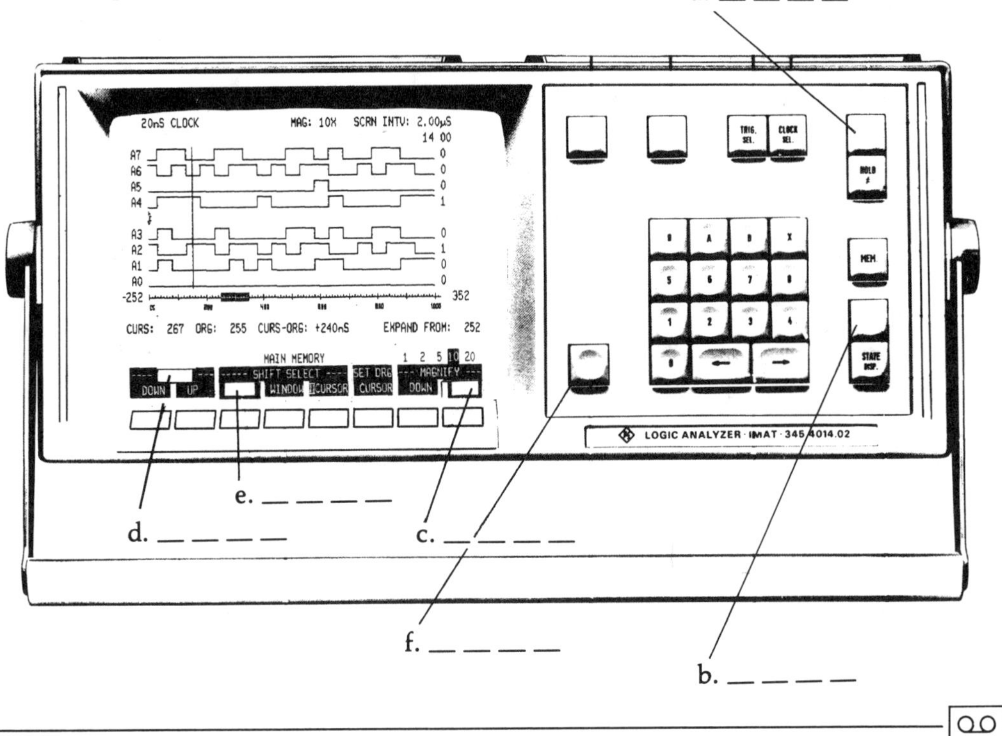

2 Presentation

The demonstrator used the following expressions to guide his audience:

TAKING STEPS
Lets's look at . . .
leave it . . .
just enter . . .
If we press this button, *we can* choose . . .
Press this button *and we*'re back to . . .

HOW TO DO IT
We can increase magnification *by pressing* . . .
We can change the position *by moving* . . .

DRAWING ATTENTION
You *can see* . . .
'll see . . .
We're *looking at* . . .

GIVING EXAMPLES
say 1.6 volts
let's say . . .

POINTS TO REMEMBER
Just one thing to remember . . .
Don't forget . . .

He used the following expressions to guide the audience to the *position* of certain features

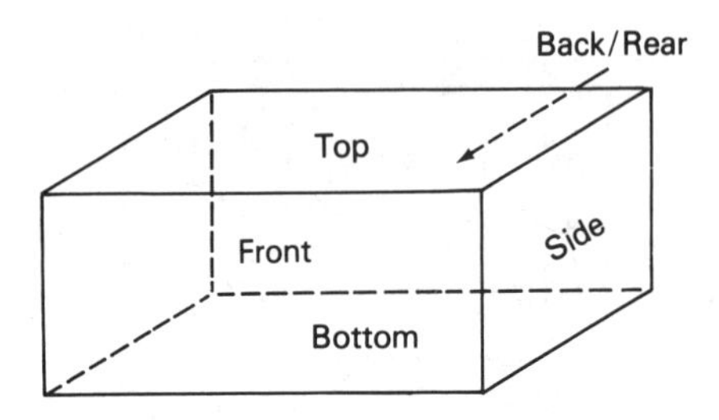

above	at the top	
the left-hand side on the left to the left	in the centre	the right-hand side on the right to the right
below	at the bottom	

Note: *on* the left of the screen

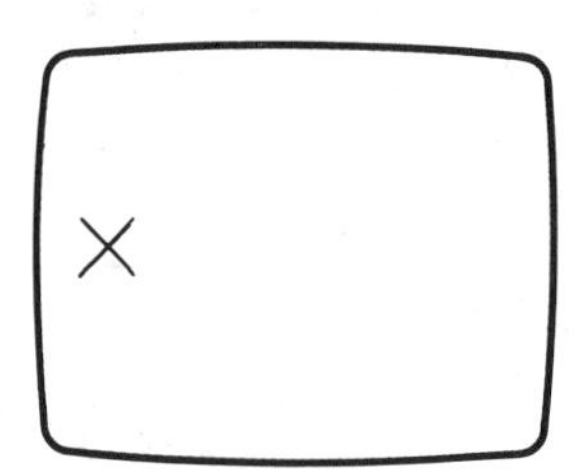

to the left of the screen

3 Controlled Practice

Complete the script of this demonstration using the diagram below:

OK, _ _ _ _ look at the back panel. You _ _ _ _ _ _ _ _ that there are two probe connectors _ _ _ _ _ _ _ _ top _ _ _ _ _ _ _ _ _ _ _ _ _ _ _ _ side. _ _ _ _ the two probe connectors, there's an IEC-bus connector. _ _ _ _ we connect this, we can transfer all the data to another system. Right, _ _ _ _ _ _ _ _ at the input and output sockets _ _ _ _ _ _ _ _ bottom of the _ _ _ _ panel. The first one _ _ _ _ _ _ _ _ left is for input of a counter and signature analyser. Next to this is the analog recorder input. _ _ _ _ _ _ _ _ centre is the video output – we can display data on another screen _ _ _ _ connecting a video terminal. _ _ _ _ _ _ _ _ right of the video output are two other sockets: one for a link line, the other for an external qualifier.

Of course, _ _ _ _ _ _ _ _ _ _ _ _ _ _ _ _ remember, _ _ _ _ forget to connect the power supply to the three pin socket _ corner!

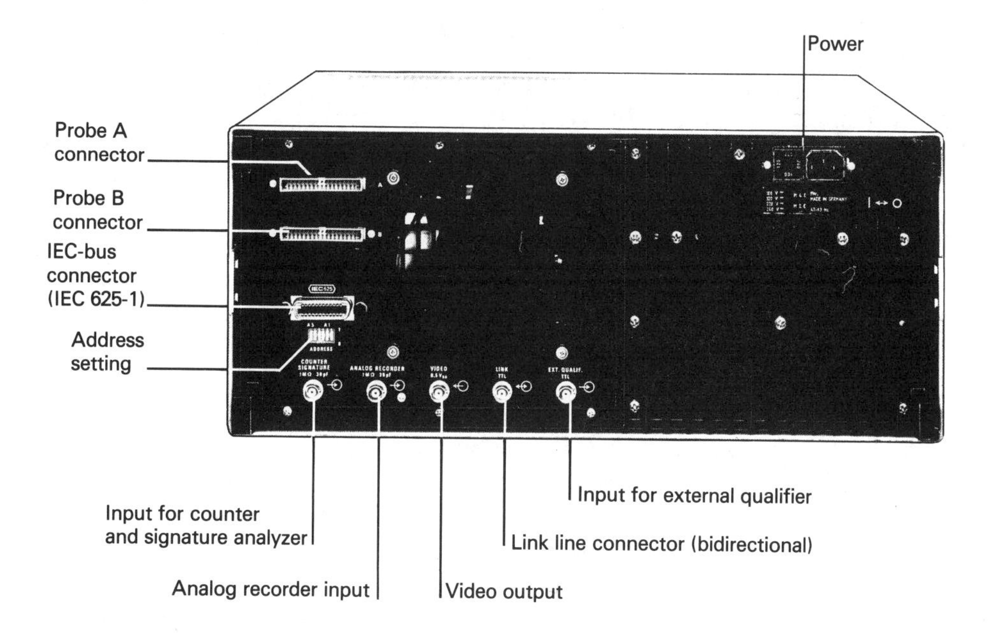

4 Transfer

Demonstrate a device in the classroom.
Suggestions for demonstration:
common household appliances: a coffee machine, a cassette recorder, an alarm clock, etc.
specialist devices: a microcomputer, a multimeter, a photocopier, etc.

5 Word Check

PROCESS VERBS
to set up — to prepare, to make ready
to enter (data) — to key in, using the data entry keys
to press — to push down on a key
to magnify/magnification — to make a picture/image bigger
to scroll up/down — to move text/image on the screen up or down

OTHER WORDS
general purpose — for most uses
a tool — an instrument, device
a mode — way of using something
frequency — the number of cycles in transmission (e.g. radio), measured in hertz
parameters — the limits for measurement
a window — a part of the screen which displays data from another part of the computer's memory

UNIT 8 Setting up your new Computer

(instructions – imperatives)

Introduction

This unit describes the steps to be followed to set up a personal computer. The equipment provided in the package is:

- the computer
- a power supply unit
- an aerial lead
- a micro-disk drive
- a micro-disk drive lead

1 Listening

You are going to hear an extract from the introductory tape supplied with the QD Personal Computer. The first section explains how to set up the system. A VDU is not provided in the package, and so the user needs to use a TV set. As you listen, number the steps in the boxes in the diagram. There are 12 steps to number. The first one has been done for you.

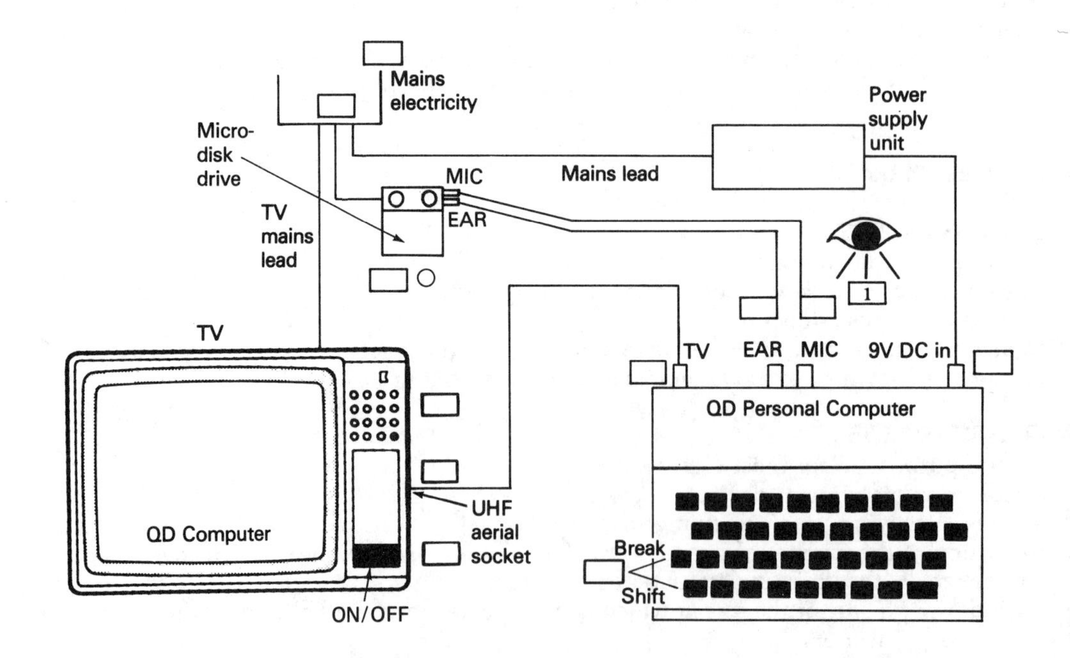

2 Presentation

On the tape you heard various instructions about how to set up the QD Personal Computer. These instructions were either in the positive or negative. Look at the examples below.

POSITIVE

Look at the back of the computer
Find the aerial lead
Connect one end of the aerial lead
Take the micro-disk drive
Do the same with the other end
Tune in your TV
Turn it *on*
Press a button
Use the tuning mechanism
Insert the 'Welcome' disk

NEGATIVE

Don't use the VHF one (aerial socket)
Don't worry about which end you use
Don't forget to turn the volume down
Don't use force
Don't try to repair it yourself

These forms of the verb are called *imperatives*.
The positive imperative is the verb stem, e.g. look, find, etc.
The negative imperative is *don't* + the verb stem, e.g. don't use, don't try, etc.
In English the imperatives have only one form. It is the same form if you give instructions to one person or to many people.

3 Controlled Practice

Now use the diagram below to give instructions. You must complete the sentences, using the verbs given. If you see an X in the diagram, you must use a negative imperative.

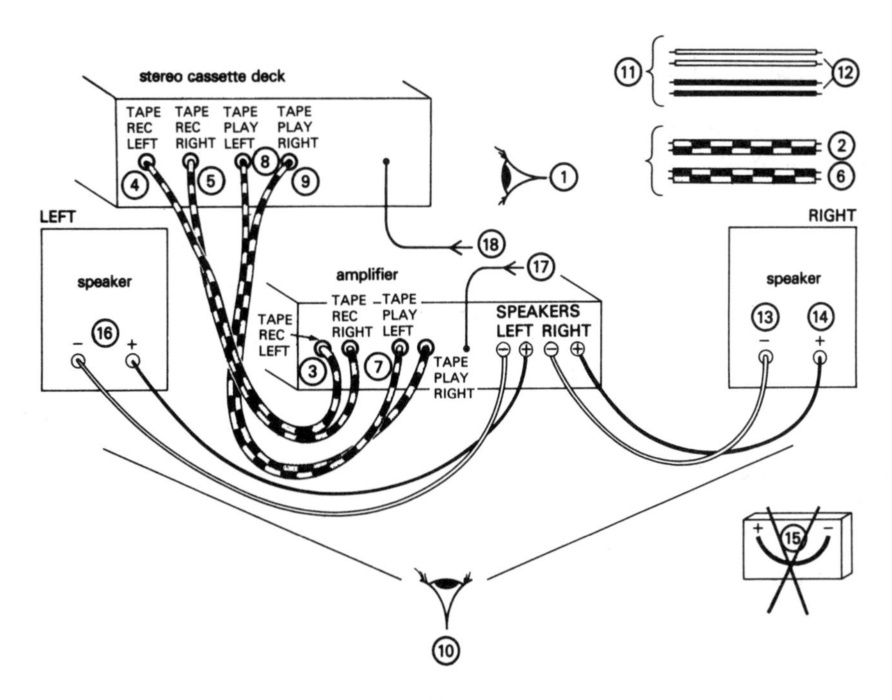

Choose from the following verbs:

take, push, follow, connect, look, do, insert, find, plug, switch

Instructions for setting up a hi-fi system.

1. _ _ _ _ at the back of the cassette deck.
2. _ _ _ _ one pair of striped leads.
3. _ _ _ _ one end of a striped lead into TAPE REC LEFT on the amplifier.
4. _ _ _ _ the other end into TAPE REC LEFT on the tape deck.
5. _ _ _ _ the same with the other lead to connect TAPE REC RIGHT to TAPE REC RIGHT.
6. _ _ _ _ the other pair of striped leads.
7. _ _ _ _ one end of the striped lead into TAPE PLAY LEFT on the amplifier.
8. _ _ _ _ the other end into TAPE PLAY LEFT on the tape deck.
9. _ _ _ _ the same sequence with the other lead to connect TAPE PLAY RIGHT to TAPE PLAY RIGHT.
10. _ _ _ _ at the back of the loudspeakers.
11. _ _ _ _ the 4 loudspeaker leads.
12. _ _ _ _ one white and one black lead.
13. _ _ _ _ one end of the white lead to the negative socket on the right speaker channel and the other to the negative socket on the right amplifier channel.
14. _ _ _ _ the black lead to the positive channel on the right speaker and the positive socket on the right amplifier channel.
15. _ _ _ _ connect positive to negative.
16. _ _ _ _ the same with the left channel and the left speaker.
17. _ _ _ _ the amplifier into the mains electricity.
18. _ _ _ _ the cassette deck into the mains electricity.

4 Transfer

PAIRWORK

Student B: Turn to Key Section

Student A: Below are the instructions for inserting fanfolded paper into a printer. Instruct your partner how to do it.

1.

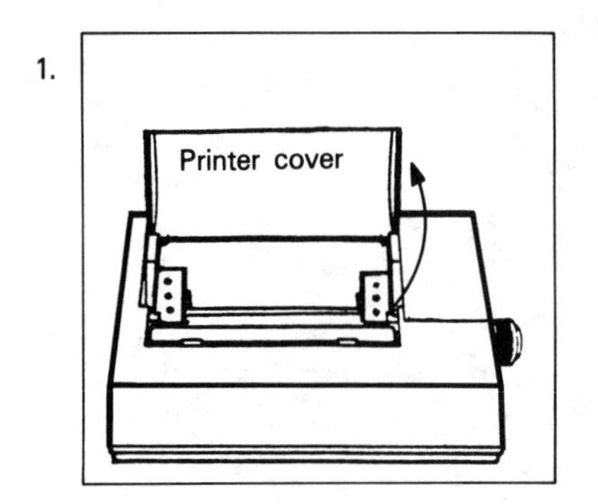

Open the printer cover

2.

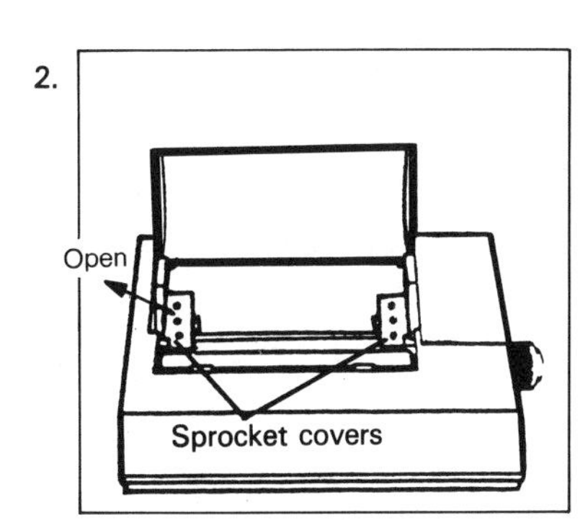

Open the left sprocket cover

3.

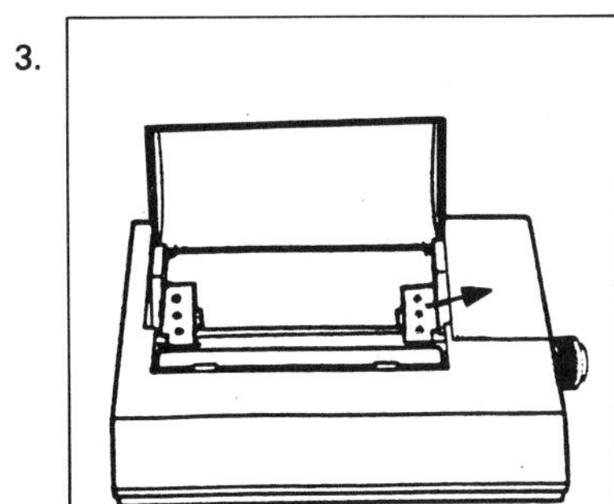

Open the right sprocket cover

4.

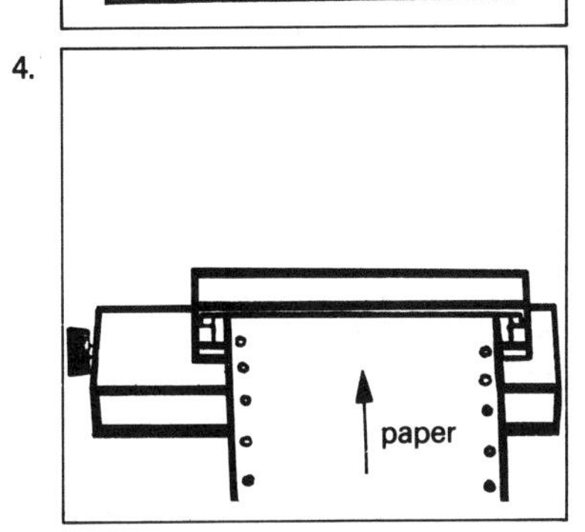

Push the paper into the back of the printer

5.

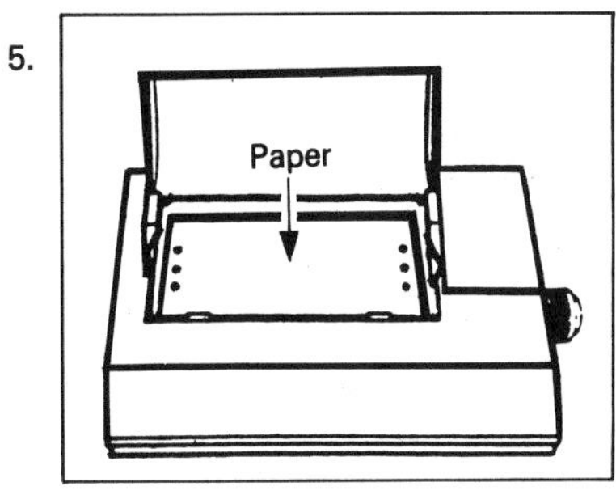

Match the holes on the paper to the sprockets

6.

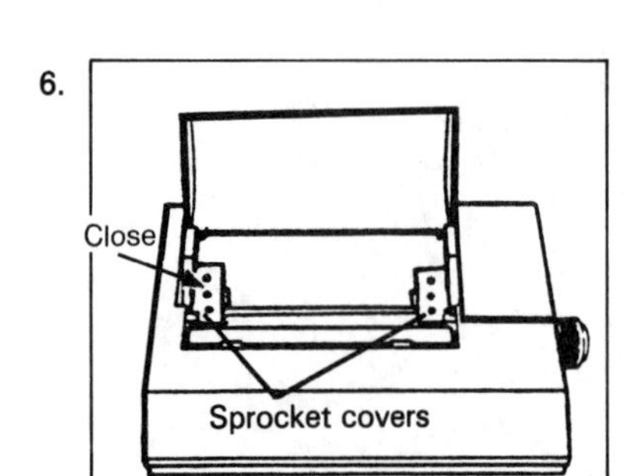

Close the sprocket covers

7.

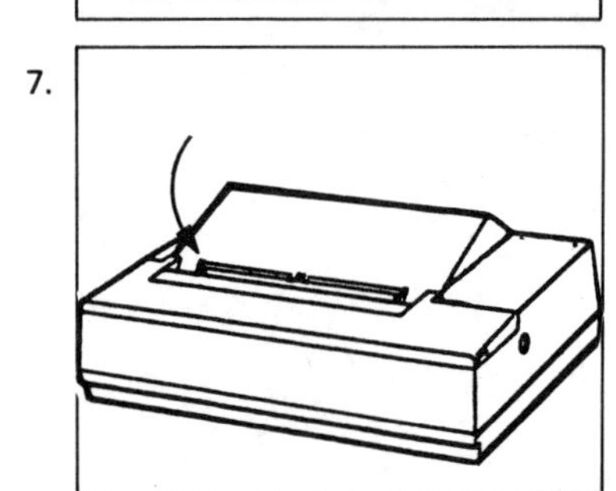

Close the printer cover

8.

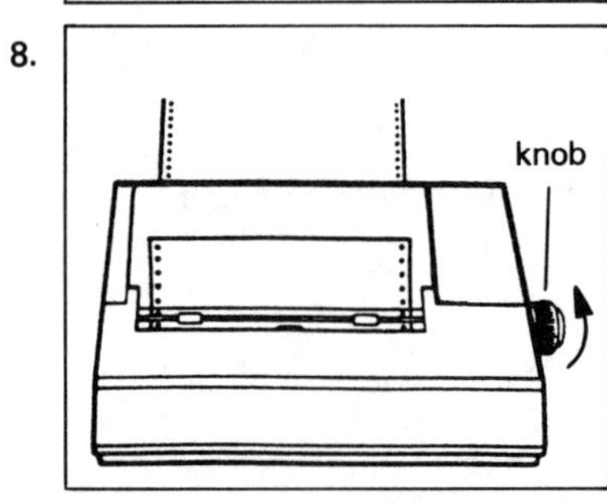

Turn the paper feed knob to advance the paper

9.

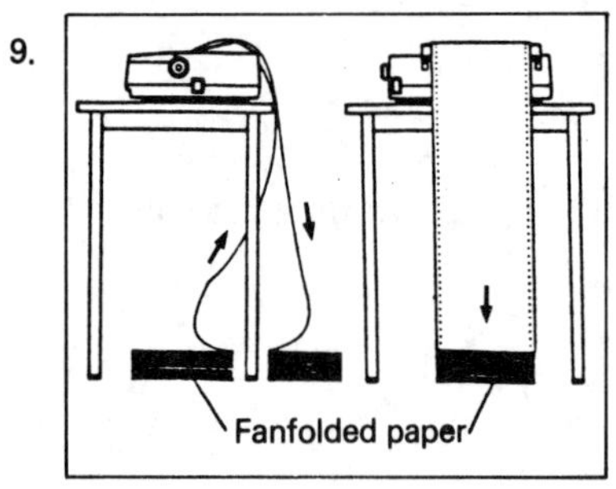

Put the paper under the printer

5 Word Check

EQUIPMENT AND PARTS
socket – opening into which to fit a plug
cable – a set of electrical wires
label – piece of paper or other material with information on it

SETTING UP THE EQUIPMENT
to connect – to link an electrical device to the power supply
to plug . . . into – to link an electrical device to the power supply
to tune in – to set a TV to receive a TV or computer signal
to insert – to put in

UNIT 9 Building a House – the Requirements

(probability and forecasting)

Introduction

This unit deals with the protection needed by a house in Britain against the *elements*, the *environment* and other *risks*. In particular the speaker considers *solar radiation*, rain, winds, noise, *damp*, fire, *heat loss* and snow.

1 Listening [OO]

In the following extract a structural engineer presents the factors which need to be considered in house building in Britain. The measures for protection can be divided into:

certainly required
probably required
possibly required
probably not required
certainly not required

As you listen tick (√) the appropriate column in the table below to indicate the requirements of a British city house.

FUNCTIONAL REQUIREMENTS OF A BRITISH CITY HOUSE

	Certainly required	*Probably required*	*Possibly required*	*Probably not required*	*Certainly not required*
Protection against solar radiation					
Protection against rain					
Protection against high winds					
Protection against noise from outside					
Protection against noise from inside					
Protection against damp from underground					
Protection against fire from outside					
Protection against fire from inside					
Protection against heat loss					
Protection against heavy snow					

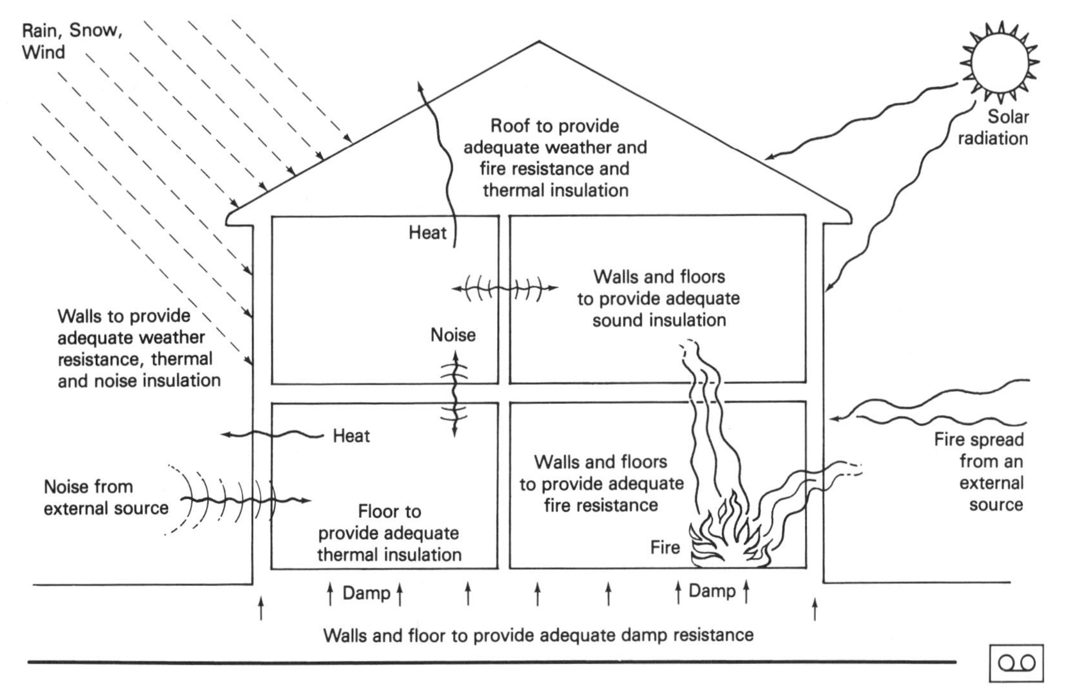

2 Presentation

In his talk the structural engineer spoke about the need for protection. Here is some of the language that he used.

	Scale of likelihood
Certainly	I am {sure / certain / positive} that a house will need protection against rain. A house is {certain / bound} to need protection against rain. A house will {certainly / definitely} need protection against rain.
Probably	It is likely that a house will need protection against noise from outside. A house is likely to need protection against noise from outside.
Possibly	A house {may / might} need protection against noise from outside.
Probably not	It is unlikely that a house will need protection against solar radiation. A house is unlikely to need protection against solar radiation.
Certainly not	A house {definitely / certainly} won't need protection against heavy snow. I am {sure / certain / positive} that a house won't need protection against heavy snow.

3 Controlled Practice

Now look at the table below and use the information given to make sentences. The tick indicates the scale of likelihood.

	Certainly	*Probably*	*Possibly*	*Probably not*	*Certainly not*
1. We hear traffic noise from outside.		√			
2. The house loses heat.	√				
3. The house becomes too hot.					√
4. Damp rises into the house.			√		
5. Rain comes through the roof.				√	
6. We feel the wind through the walls.					√
7. Fire spreads inside the house.		√			
8. The roof collapses because of snow.					√
9. We hear noise from other parts of the house.	√				
10. Fire spreads from outside the house.			√		

1. It is likely that we will hear traffic noise from outside.
   We are likely to hear traffic noise from outside.
2. ______________________________

   ____
3. ______________________________

   ____
4. ______________________________

   ____
5. ______________________________

   ____
6. ______________________________

   ____
7. ______________________________

   ____
8. ______________________________

   ____

9. ______________________________

____

10. ______________________________

____

4 Transfer

Now use the language introduced in this unit to discuss:

a the protection needed by a city house in your country

b the protection needed by a village house in your country.

5 Word check

STRUCTURE

shutter — wood or metal cover for protection outside a window
blind — wood, fabric or metal cover for protection inside a window
brick — building material made of baked clay
roof — outside cover on top of a building
double-glazed window — window with 2 sheets of glass
foundations — base of a house put deep into the ground to support the walls
solid — made of hard material
thick — opposite of thin
sloping (roof) — at an angle

CLIMATE

temperate — neither very hot nor very cold
monsoon — season of heavy rains

UNIT 10 Installing the 9450 Photocopier

(instructions – modals)

Introduction

This unit deals with the *assembly* and *installation* of the 9450 photocopier.

1 Listening

A group of newly recruited service engineers are attending a briefing course on photocopier installation. The Technical Manager is giving the basic instructions for the installation of the 9450. As you listen, tick the appropriate column in the table. DO's mean actions that the engineer must perform; DON'Ts mean actions that the engineer mustn't perform.

Instruction	*Do's*	*Don'ts*
check if the customer wants a cabinet		
presume he doesn't want a cabinet because it isn't on the order form		
try to persuade the customer to buy a cabinet		
assemble the copier on the floor		
assemble the copier in its final position		
adjust the feet so that the machine is level		
use force to attach the lid		
slide the A4 cassette holder into the lower space		
fit the tray under the A4 cassette holder		
pour 2 bottles of 5100 toner into the toner compartment		

2 Presentation

The instructions given by the Technical Manager were of 2 types: **POSITIVE** and **NEGATIVE**. Look at the examples below.

POSITIVE
You must check if he wants the cabinet.
You ought to try to persuade him . . .
You should assemble the copier . . .
You are to take out the main copier.
You are supposed to just click it into position.

NEGATIVE
You mustn't use any other toner except 9450.
You oughtn't to need any force.
You shouldn't use a surface . . .
You are not to assemble it . . .
You are not supposed to use force.

The above verbs consist of 2 groups, as follows:

1. MODALS

POSITIVE	NEGATIVE
must	mustn't
should	shouldn't
ought to	oughn't to

The modals don't change their form with the subject,
e.g. *You must* check . . .
He mustn't pour in the toner . . .
They should try to . . .
We oughtn't to use force . . .

2. to be (supposed) to
Here the verb *to be* changes with the subject,
e.g. *I am (supposed) to* assemble . . .
You are (supposed) to shake the toner . . .
Notice the position of *not* in the negative form:
He is not (supposed) to supply free copier paper.
We are not (supposed) to presume the customer doesn't want a cabinet.

3 Controlled Practice

A. Below are the steps to be followed if the 9450 jams (i.e. if paper gets stuck in the machine). Number the steps from 1 to 12 in the correct order. Some of them have been done for you.

You must push the reset button.	
You shouldn't use force when replacing the drum.	8
You oughtn't to try to pull out the copy drum before you release the locking screw.	3
You must switch the power off.	1
You ought to replace the copy drum carefully.	
You ought to check the machine is functioning.	
You should close the front panel.	10
You mustn't forget to switch the power on again.	
You should remove the jammed sheet of paper.	
You should open the front panel.	
You are not (supposed) to touch the drum surface.	
You mustn't use a sharp object to remove the sheet.	6

B. Now look at the following operator control signals for the 9450. Then complete the sentences with one word in each gap.

1. WAIT INDICATOR You __________ wait until the machine is ready to copy.
2. READY INDICATOR You __________ __________ press the copy key now.
3. ADD PAPER INDICATOR You __________ __________ to add paper.
4. ADD TONER INDICATOR You __________ use the machine before you add some toner.
5. MISFEED INDICATOR You __________ __________ try to press the copy key as the machine is jammed.

4 Transfer

PAIRWORK

Student B: Turn to the Key Section.

Student A: You have just bought a video cassette recorder. When you arrive home, you find that the installation and operating instructions are missing. You want to set up the machine immediately, and so you telephone the shop and ask for the Technical Manager.
You ask him to tell you the steps to follow to set up the machine.
As the Manager gives you the instructions, number the pictures from 1 to 12 below, and make short notes in the space.

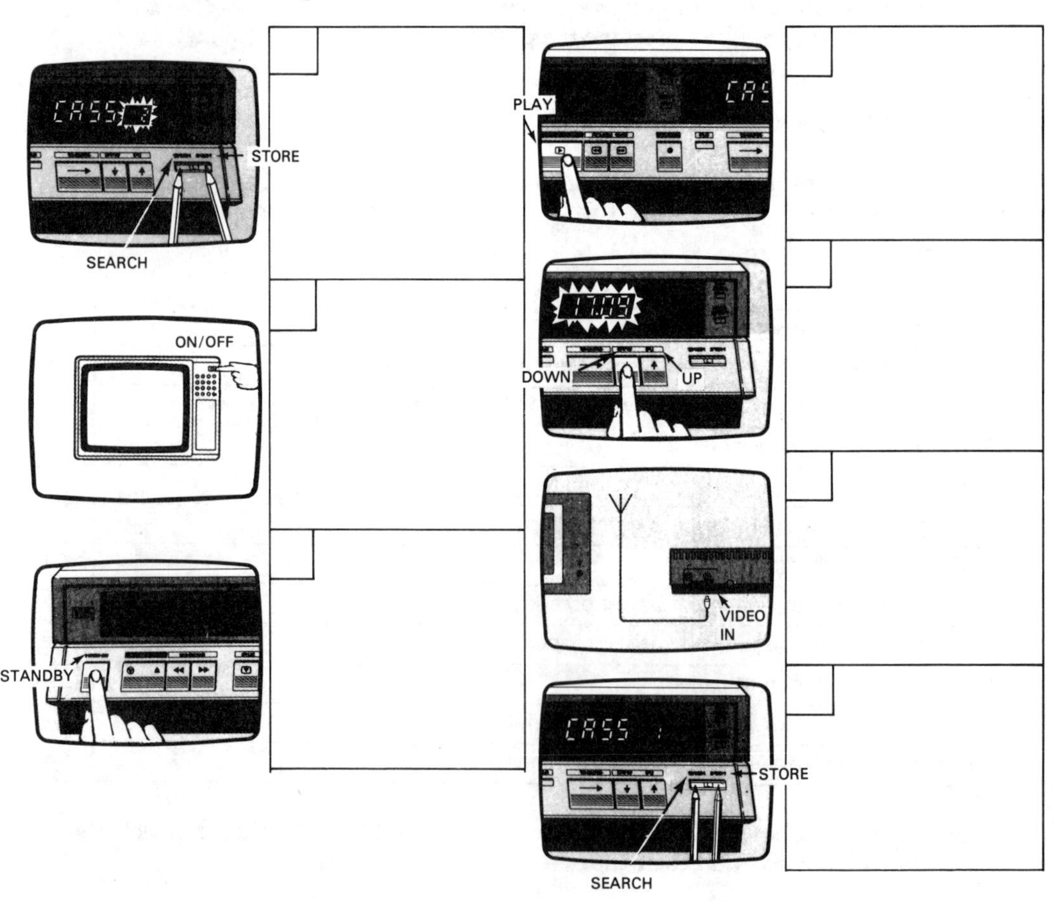

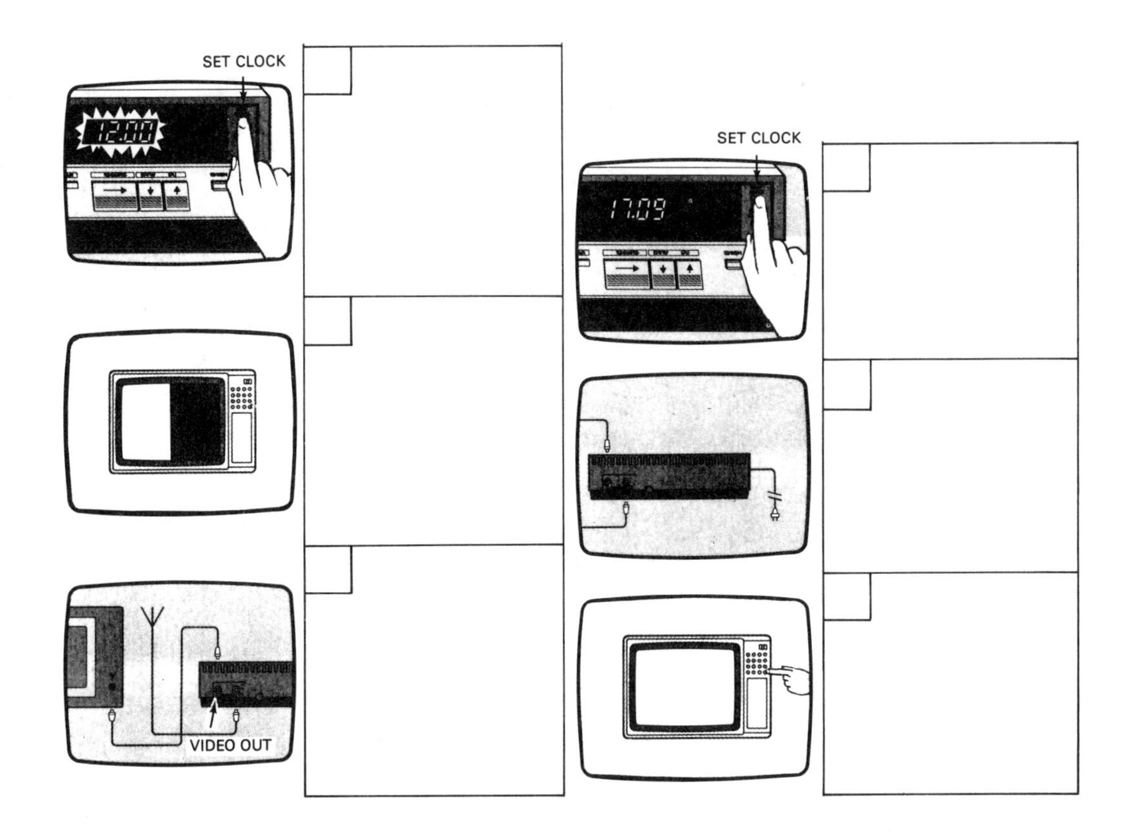

After you have listened to the instructions and numbered the pictures, use your notes to repeat to your partner what you must do, so that you are sure that you have understood correctly.

5 Word Check

EQUIPMENT AND PARTS
cabinet — piece of (office) furniture on which to place an object, e.g. a photocopier, or in which to store things
lid — part that covers the top of a photocopier and can be lifted up
paper cassette holder — container for paper in or on photocopier
copy tray — container into which finished photocopies go
toner — powder used in photocopying process to give black image

ASSEMBLY AND INSTALLATION
to assemble — to fix parts together
to adjust — to change (level, height, position, etc.) by a small amount
to attach — to fasten (one thing to another)
to slide in — to move (something) smoothly into its correct position
to shake — to move something (e.g. a bottle) quickly up and down
to pour — to transfer a liquid or powder from one container into another

UNIT 11 Printing Processes

(process description – past, present and future tenses)

Introduction

This unit deals with the modernisation of printing processes. Traditional processes were largely manual. The pages of a newspaper were *cast* (formed) in metal. Modern methods do not use metal. They use computers, photographic processes and plastic *plates* (newspapers are printed by pressing plates against paper).

1 Listening

You are going to hear the editor of a newspaper briefly reviewing the development in printing processes. As you listen, complete the flow charts below:

2 Presentation

During his talk, the editor used both ACTIVE and PASSIVE forms in **PAST, PRESENT** and **FUTURE**:

	ACTIVE	PASSIVE
PAST	The journalist typ*ed*	The story *was cast* (singular) The pages *were made* (plural)
PRESENT	The journalist type*s* The compositors type	The type *is set* Stories *are retyped*
FUTURE	The journalist *will type*	Plates *will be made*

3 Controlled Practice

Use the information in the flow charts to complete the sentences below:

A. **Past**

1. The journalist ____ the story on ____ _____ _____.
2. The story _____ ____ in metal by ____ _____.
3. The metal type _____ _____ _____ _____.
4. The pages _____ _____ _____ _____ _____ by _____ _____.
5. The complete pages _____ _____ in metal ready for _____ _____ _____.

B. **Present**

1. The journalist _____ _____ _____ _____ _____.
2. The compositor _____ _____ _____ _____ the main computer.
3. The computer _____ _____ _____ _____ onto paper.
4. Individual articles _____ _____ on to the page.
5. The pages _____ _____ and plastic plates _____ _____ for the press.

C. **Future**

1. The journalist _____ _____ _____ _____ _____.
2. The story _____ _____ _____ _____ _____.
3. The computer _____ _____ _____ _____ _____.
4. The pages _____ _____ _____ and _____ _____ on _____ by _____ _____.
5. The computer _____ _____ the pages and plastic plates _____ _____ _____ _____ for _____.

4 Transfer

GROUP WORK

Present the changes in technology for the following:

1. Telecommunications (telegraph — telephone — videophone)
2. Car manufacture (manual assembly — mass assembly lines — robotic and computerized)
3. Energy (wood — coal — oil/gas — nuclear — alternative energy)

And any others you would like to present.

5 Word Check

PROCESS VERBS

to type — to use a typewriter

to make up (a page) — to put the stories/articles in the right order/position

to set (type) — to cast letters into metal or directly onto paper photographically
to paste — to glue/stick the stories/articles onto the page
to design — see 'to make up'

WORKERS
a journalist — person who writes the stories/articles
a compositor — a technician who casts the letters in metal or sets type photographically on paper
a printer — a technician who works on the printing press

EQUIPMENT
a manual typewriter — works by hand, no electricity
an electronic typewriter — is driven by electricity and controlled by a microchip
a VDU/Computer terminal — Visual Display Unit, consisting of a screen and a keyboard
printing press — equipment for printing a newspaper, magazine, etc.

UNIT 12 Energy

(increase and decrease)

Introduction

This unit deals with the *trends* and *statistics* of electricity *generation*.

1 Listening

Robin Coates is the head of an electricity region. In this presentation he is talking about the total capacity of the electricity generating plants in his region. He describes how trends in electricity requirements have changed. As you listen, draw a graph to show these trends.

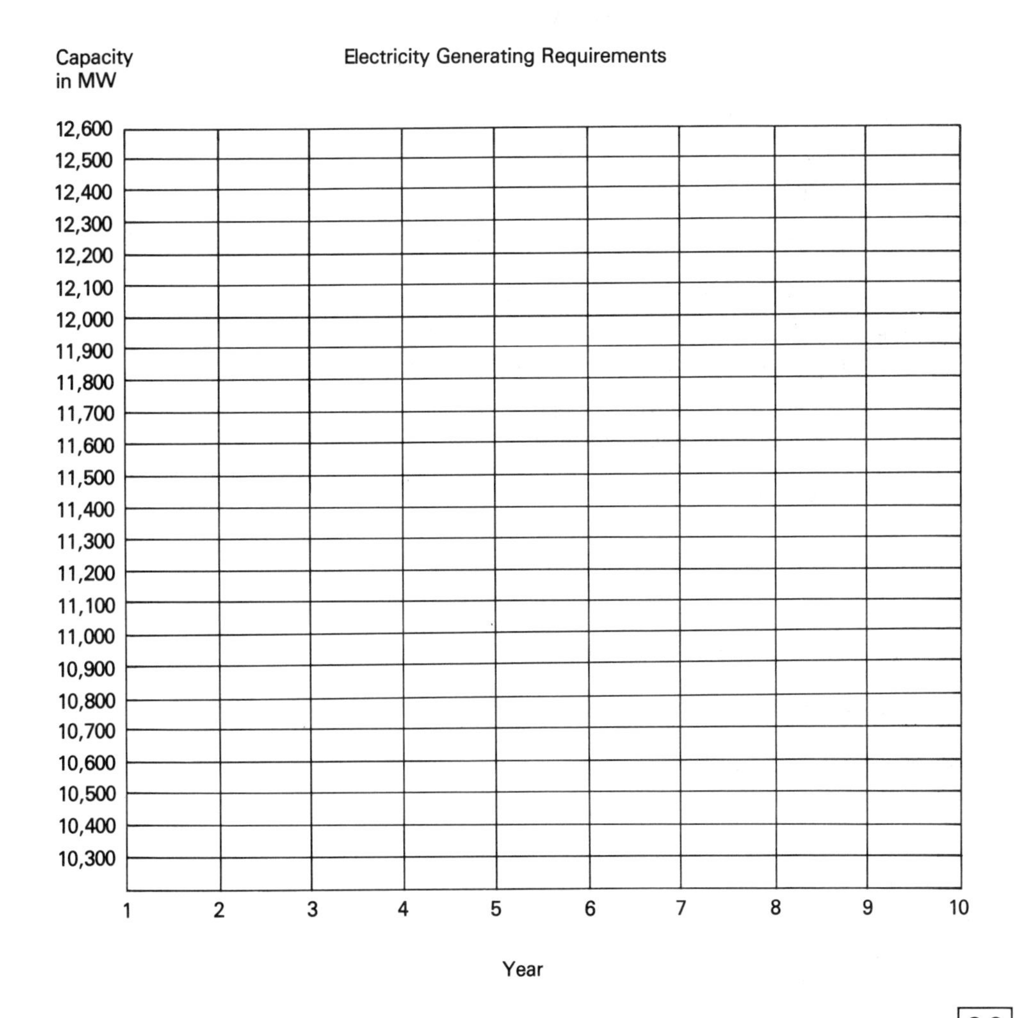

2 Presentation

In his talk Robin Coates talked about increases and decreases in generating capacity. Below you will find some of the language he used.

a. Increase

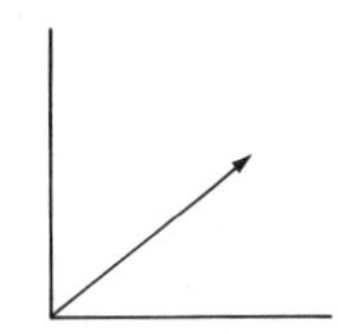

verb phrases
to increase by/to
to rise by/to
to go up by/to

noun phrases
an increase of/to
a rise of/to

b. Decrease

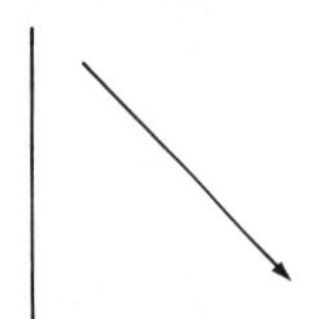

verb phrases
to decrease by/to
to fall by/to
to drop by/to
to go down by/to

noun phrases
a decrease of/to
a fall of/to
a drop of/to

Compare the following 2 sentences:

i) In year 5 capacity increased by 200 MW.
ii) In year 5 capacity increased to 12200 MW.

In sentence i) we use 'by' to focus on the difference between the previous level and the level in the 5th year. In sentence ii) we use 'to' to focus on the level in the 5th year.
Now compare the following 2 sentences:

i) In the 7th year we saw a decrease of 400 MW.
ii) In the 7th year we saw a decrease to 12000 MW.

In sentence i) we use 'of' to focus on the difference between the previous level and the level in year 7. In sentence ii) we use 'to' to focus on the level in year 7.

c. Other expressions

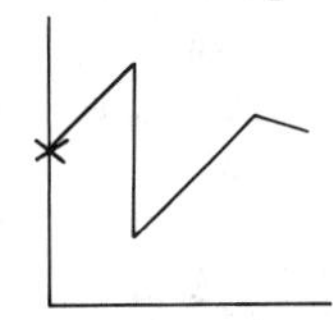

to stand at

We use this phrase to focus on a particular point, before we mention the trends or movements,
e.g. In the 1st year electricity capacity stood at 10900 MW.

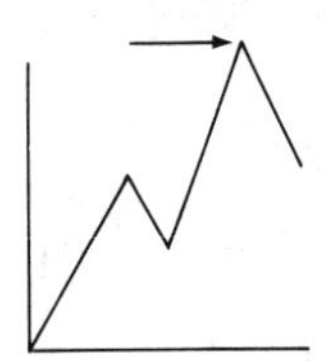

to reach a peak of

e.g. In the 6th year the generating capacity in our region reached a peak of 12400 MW.

to remain constant at

e.g. In year 3 capacity remained constant at 11200 MW.

3 Controlled Practice

Look at the graph below and complete the sentences by writing one word or number in each gap.

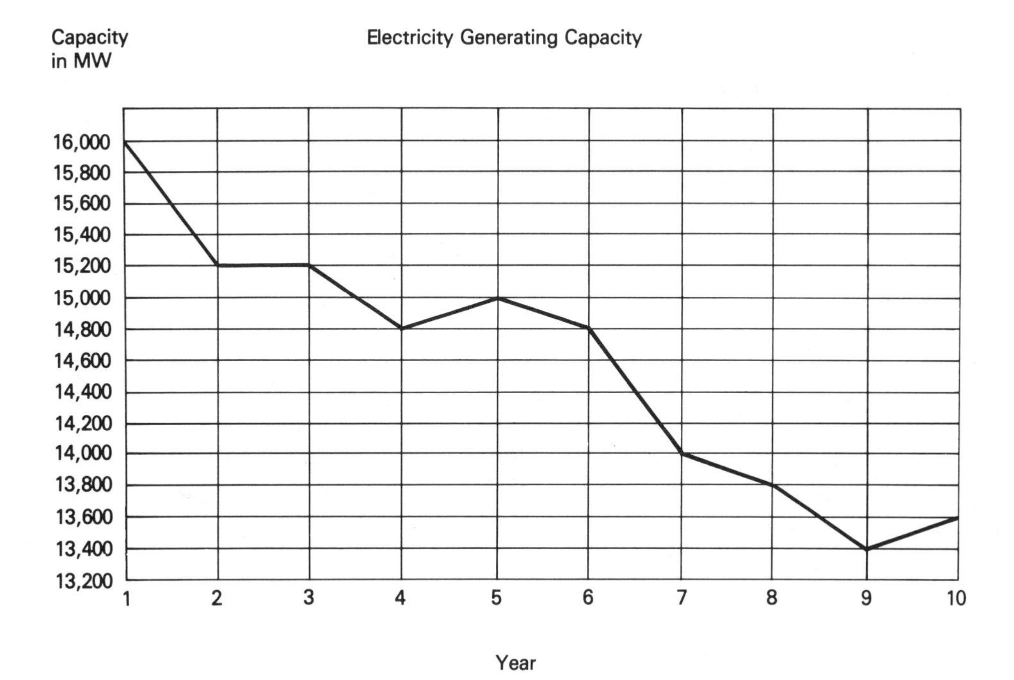

1. In the 1st year capacity ________ ________ 16000 MW.
2. In year 2 it ________ by 800 MW.
3. In the 3rd year it ________ constant ________ 15200 MW.
4. Then in the 4th year it ________ ________ 14800 MW.
5. And in year 5 we saw a small increase ________ 200 MW.
6. But in the 6th year capacity ________ to 14800 MW.
7. This trend continued, and in the 7th year capacity ________ substantially ________ 800 MW.
8. There was a further ________ ________ 200 MW in the 8th year.
9. And in year 9 capacity ________ ________ ________ 13400 MW.
10. But in year 10 capacity showed a small ________ ________ 200 MW.

4 Transfer

PAIRWORK

Student B: Turn to Key Section.

Student A: Look at the completed graph for the total electricity production of two countries. Then describe the trends to your partner, who will complete the graph in the Key Section.

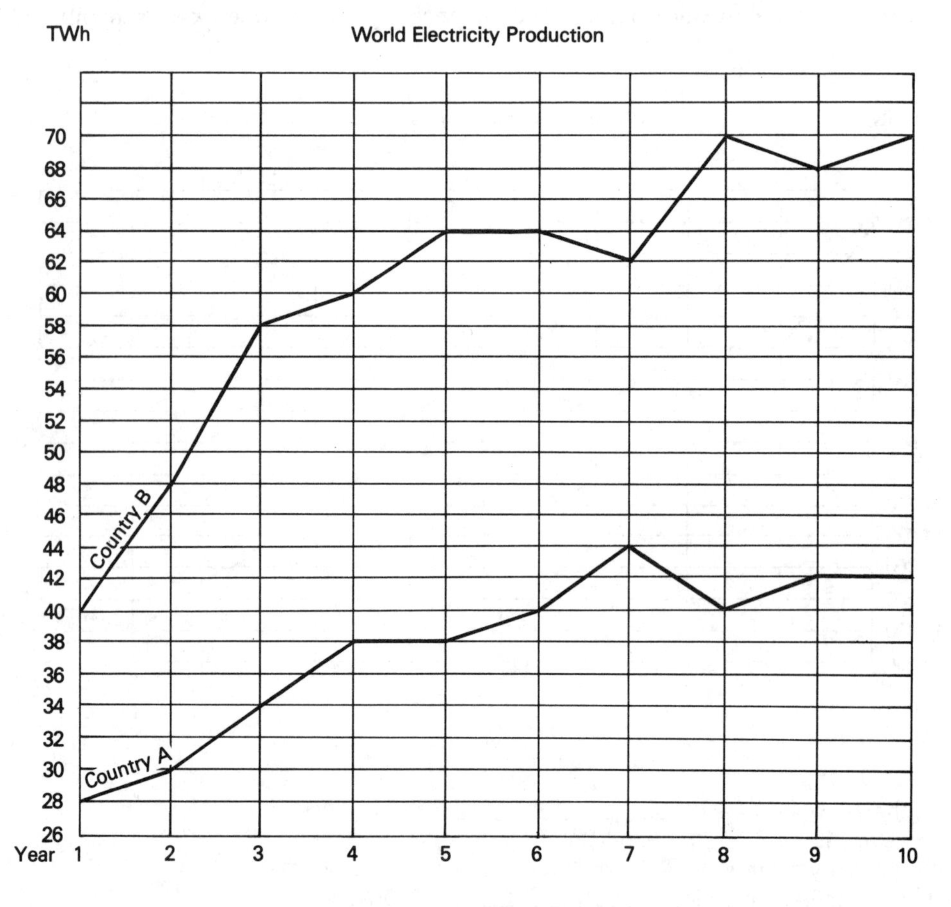

Student A: Now complete the graph below using the information that your partner will give you about world electricity production. After you have completed your graph, compare it with your partner's.

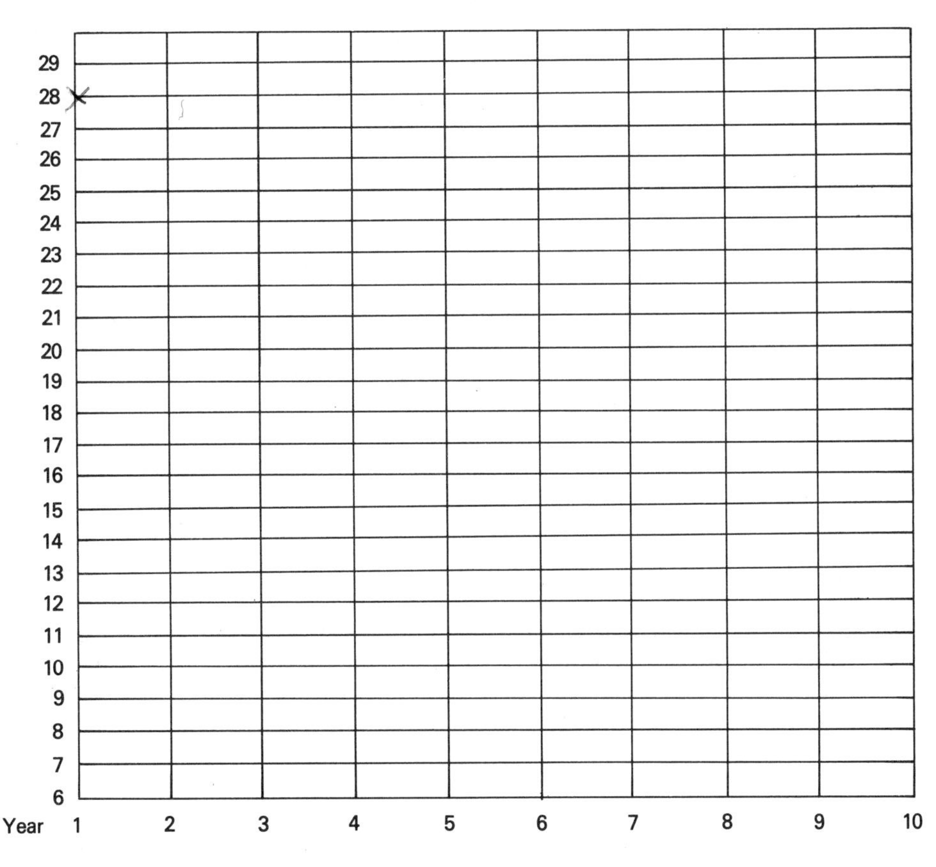

TWh = Terra Watt hours

5 Word Check

DEGREE OF CHANGE
steady — regular
moderate — medium
substantial — at a high level
dramatic — at a very high level

OTHER WORDS
capacity — maximum amount or quantity
to represent — to show
to conserve (also conservation) — to use carefully without waste
improvement — a better level, condition, etc.
to reduce — to decrease the amount or quantity
consumption — total use
to register — to show
to suffer — to experience difficulty
recession — period of low economic activity

UNIT 13 Microchip Manufacture

(process description – passives/sequencing)

Introduction

This unit deals with microchip manufacture. It describes how *wafers* (thin slices) of silicon are covered with *photoresist* (a plastic substance which reacts to light). The image of the circuit is printed on the wafer using a photographic *mask* (a film plate through which light can pass in places). The light causes parts of the photoresist to harden. The other parts are *removed* (taken off).

1 Listening

You are going to hear the production manager of a microchip manufacturer talking to a group of visitors. Before he takes them round the factory, he explains the process using the flow chart below.
As you listen, complete the labels by inserting the correct verb.

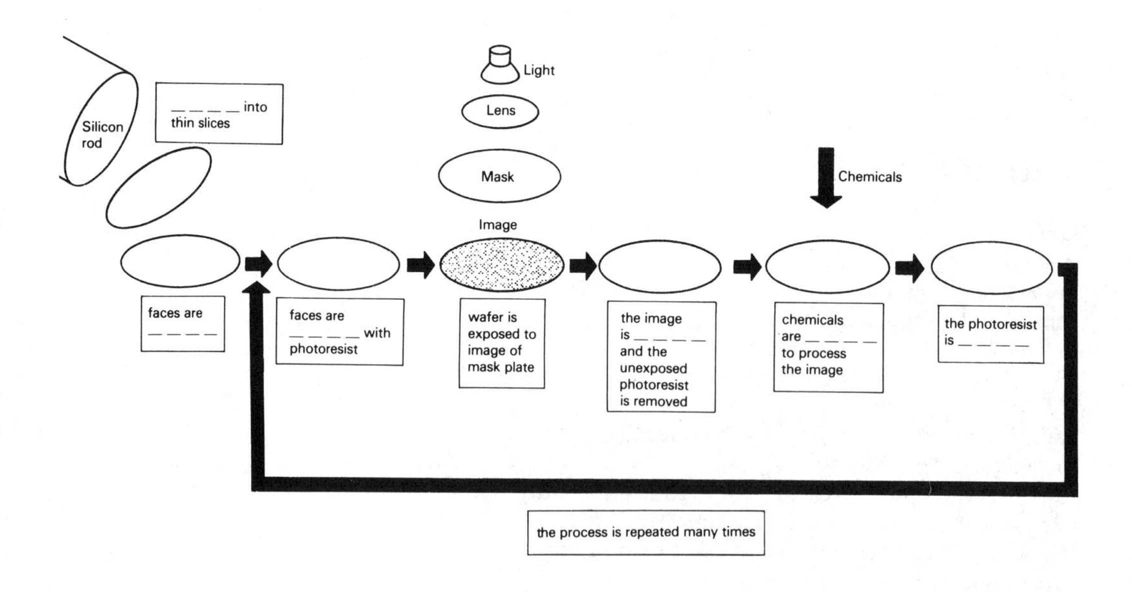

2 Presentation

During his talk, the production manager used:

A. The *Passive* form of many verbs

We use the passive when we are more interested in the OBJECT (the chip, the rod, etc.) than the SUBJECT (the machine, the operator, etc.), e.g.

The rod *is cut* into thin slices (PASSIVE)

is more appropriate than

Somebody/A machine *cuts* the rod into thin slices (ACTIVE)

Form of the passive:

It is constructed from the verb 'to be' + the past participle

e.g. the rod *is cut* (Present Tense, singular)
the wafers *are produced* (Present Tense, plural)

B. *Sequence* markers

We use these so that the listener can easily follow the *order* of the process from *first* to *last* step.

e.g. First/To start with
Then . . .
Next . . .
Before enter*ing* . . .
After remov*ing* . . .
Having remov*ed* . . .
Once . . .
Now, . . .
At the next stage . . .
Finally, . . .

3 Controlled Practice

A. Improve these sentences by changing them into the PASSIVE:

1. We cut the rods into thin slices.

_ _

2. A machine polishes the faces of the wafer.

_ _

3. We cover the faces with photoresist.

_ _

4. We expose the wafer to the image of a mask plate.

_ _

5. We develop the image.

_ _

6. A solvent removes the unexposed photoresist.

— —

7. We apply chemicals to process the wafer.

— —

8. We remove the photoresist.

— —

B. Use the flow chart illustrating the manufacturing process for printed circuit boards (pcbs) to complete the written description. Insert appropriate *sequence* markers:

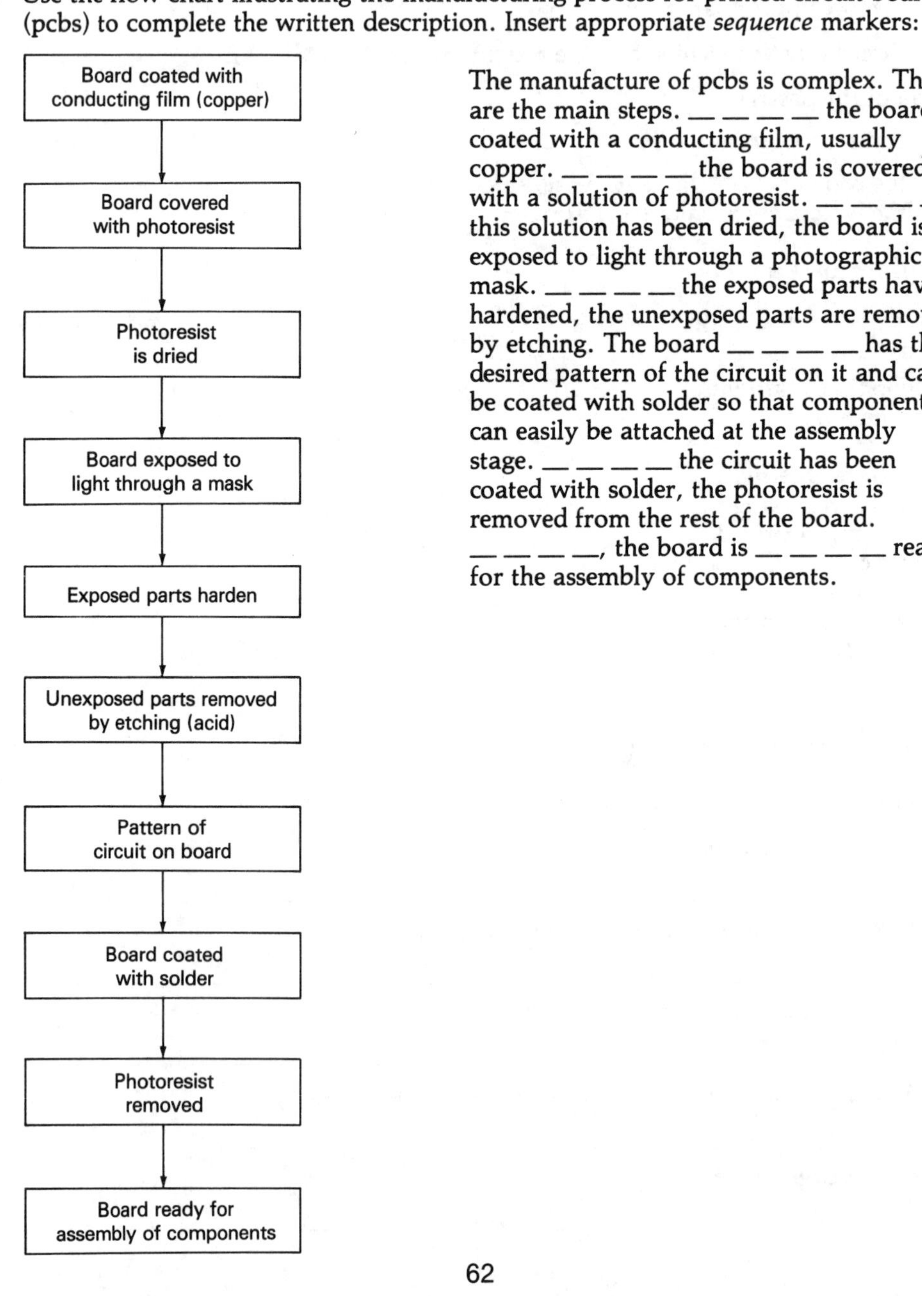

The manufacture of pcbs is complex. These are the main steps. — — — — the board is coated with a conducting film, usually copper. — — — — the board is covered with a solution of photoresist. — — — — this solution has been dried, the board is exposed to light through a photographic mask. — — — — the exposed parts have hardened, the unexposed parts are removed by etching. The board — — — — has the desired pattern of the circuit on it and can be coated with solder so that components can easily be attached at the assembly stage. — — — — the circuit has been coated with solder, the photoresist is removed from the rest of the board. — — — —, the board is — — — — ready for the assembly of components.

4 Transfer

PAIRWORK

Student A: Ask Student B questions in order to complete the flow chart below – the second part of the manufacturing process of microchips.

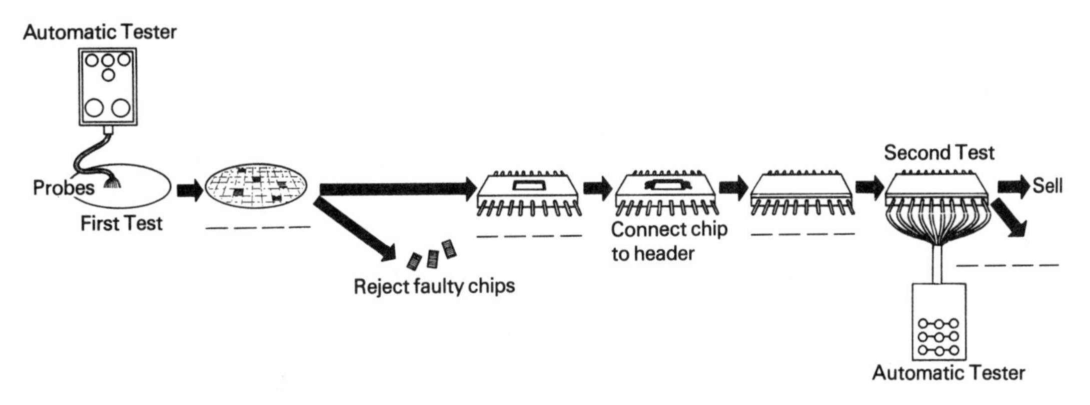

5 Word Check

PROCESS VERBS

to polish – to make a surface shiney (like a mirror)
to expose – to let light contact film
to develop – to process exposed film
to harden – to become hard
to apply – to add, to use
to mount – to fix a part onto a main structure
to coat – to cover with film, paint, etc.
to etch – to cut lines in a surface
to reject – to fail, to *not* accept
to seal – to cover/protect so that air, dirt cannot enter

OTHER WORDS

solder – an alloy (often Sn + Pb) used to join two pieces of metal
a probe – an electric lead which is connected to a circuit for testing
a header – a main structure to which the chip is mounted

UNIT 14 VCR Recording

(obligation – past modals and conditional III)

Introduction

This unit describes the steps to be followed to make *unattended* recordings of TV programmes with a video cassette recorder.

1 Listening

A customer has just bought a video cassette recorder (VCR). When he arrives home, he finds that one part of the instruction manual is missing. The section that is missing explains how to set up the VCR for unattended recording of TV programmes. The customer tries to set up the machine, but is not successful. So he rings up the shop and asks to speak to a Technical Assistant. The Technical Assistant tells him what he should have done (but didn't do). As you listen, draw a line to connect the step which the customer should have taken with the picture showing the result that would have happened.

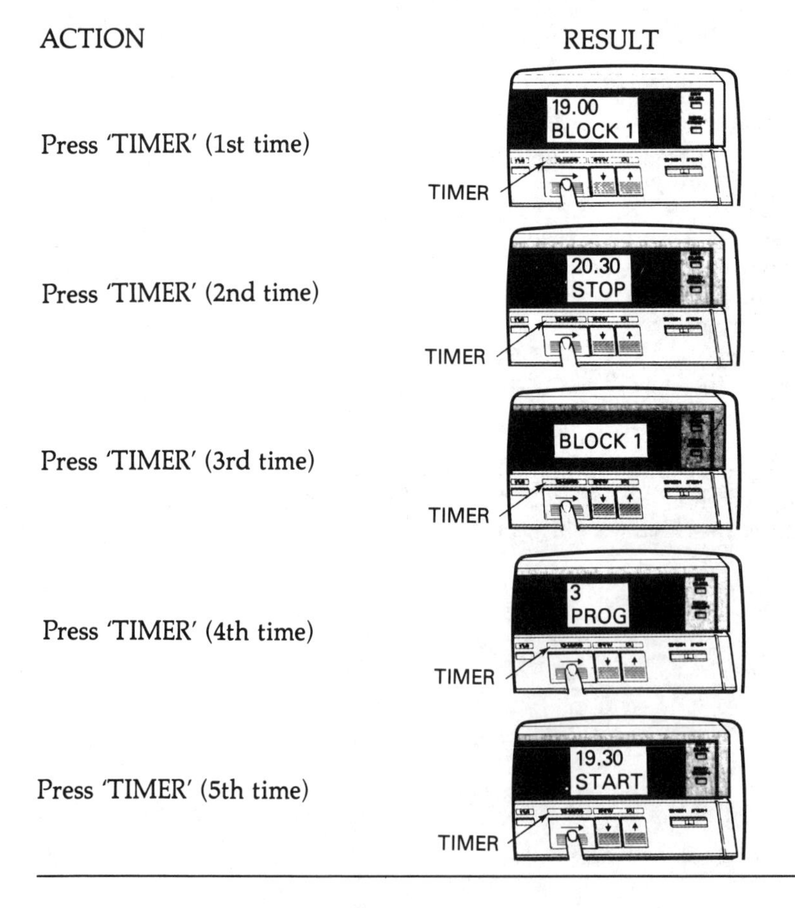

2 Presentation

In the telephone conversation the Technical Assistant told the customer what he

should / *ought to* } *have done,* (but didn't)

shouldn't / *oughtn't to* } *have done,* (but did).

e.g. i) You *should have pressed* TIMER first.
This means 'It was necessary for you to press TIMER first'.

ii) You *ought to have pressed* TIMER again.
This means 'It was necessary for you to press TIMER again'.

iii) You *shouldn't have pressed* STANDBY first.
This means 'It was wrong for you to press STANDBY first'.

iv) I *oughtn't to have pressed* STANDBY then.
This means 'It was wrong for me to press STANDBY then'.

The Technical Assistant also told the customer what would have happened if he had taken the correct action. (Unfortunately he didn't take the correct action, so the desired result did not happen).

e.g. i) If you *had pressed* TIMER, you *would have seen* BLOCK 1 on the VCR display.
(But he didn't press TIMER, and therefore he didn't see __ __ __ __)

ii) If you *had done* this, you *could have selected* the starting time.
(But he didn't do this, and therefore he couldn't select __ __ __ __)

3 Controlled Practice

Now look at the table below. In the first column you see the action that you should have taken. In the second column you see the result that would have happened if you had taken this action. For each line of information make 2 sentences. The first one has been done for you.

Action	*Result*
1. I/read the operating manual	I/see the section on voltage regulation
2. I/read the section on voltage regulation	I/not set it to 110 volts
3. I/not leave it on 110 volts	it/work
4. I/not open the case	I/not get an electric shock
5. I/switch off the power first	I/still have my VCR
Moral	
6. I/not try to be a technician	I/not lose a lot of money

1a. I should/ought to have read the operating manual.
 b. If I had read the operating manual, I would have seen the section on voltage regulation.

2a. ______________________________
 b. ______________________________

3a. ______________________________
 b. ______________________________

4a. ______________________________
 b. ______________________________

5a. ______________________________
 b. ______________________________

6a. ______________________________
 b. ______________________________

4 Transfer

PAIRWORK

Student B: Turn to Key Section

Student A: You are the Project Manager responsible for the implementation of a plan to manufacture and transport chemicals.

You designed the critical path analysis (activity flowchart) below for the activities involved. The lowest number (1) shows the activity that should have been done first. (2) shows the activity that should have been done next, etc. If the numbers are the same, the activities should have been done at the same time.

Specify Truck Chassis 1	Requisition Chassis 2	Order Chassis 4	Supplier Manufactures Chassis 7	Supplier Delivers Chassis 10		
Specify Instrument Design 2	Requisition Instruments 3	Order Instruments 5	Supplier Manufactures Instruments 9	Supplier Delivers Instruments 11		
Order Sample Chemicals 3	Check Quality of Chemicals 6	Approve Chemicals 8	Place First Order for Chemicals 10	Supplier Delivers Chemicals 11	Store Chemicals 13	
			Start Plant Tests 10	Review Tests 12	Approve Start-Up 14	Start Up Plant 15

While investigating the development of the project, you find that your Project Co-ordinator (**Student B**) has modified the plan. Check the order in which he implemented the steps, and then tell him what he should have done. You start as follows:

A: So what did you do first?
B: First I specified the truck chassis, and at the same time I specified the instrument design.
A: You shouldn't have specified the instrument design first.
You ought to have specified only the truck chassis.
And what did you do next?

5 Word Check

to connect — to link 2 parts together
to set up — to assemble
to figure out — to understand
message — visual information on the VCR
display — part of the VCR that gives visual information
manual — book of instructions

UNIT 15 Data Communications

(tenses – present)

Introduction

This unit deals with the transmission of data between computers. Data communication networks use *modems* to change data (bits of information) into sound signals (sound waves). These signals can then be sent down a normal telephone *line*. When they reach the other end, another modem changes them back into data. The receiving computer can then *store* the information on *file* (a disk will usually hold many files – each file will hold a certain type of information) or display the information on screen.

1 Listening

You are going to hear the administrative manager and communications engineer of an international firm discussing a data communication problem. This firm sends documents (reports, contracts, etc.) to their clients by connecting up computers. As you listen, label the diagrams below:

EXISTING CONFIGURATION

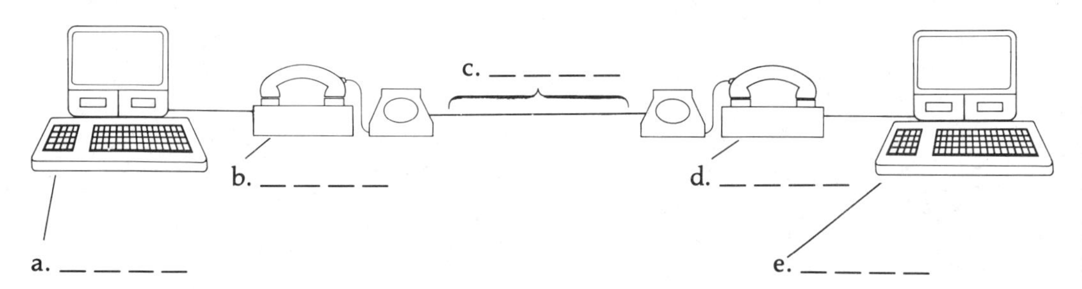

NEW CONFIGURATION (proposed)

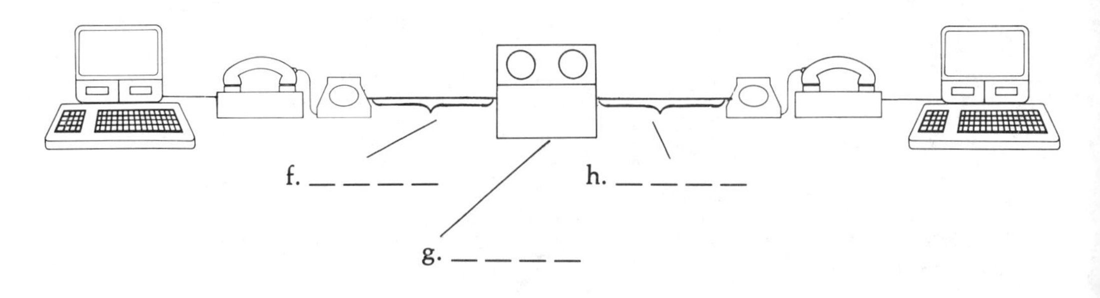

2 Presentation

During the discussion, the Communications Engineer used two tenses:

A. **Present Continuous** — to talk about how the system *is working at the moment*
   e.g. We*'re using* a modem
   We*'re having* problems
   The files *are being* transmitted
   They*'re* not *arriving*
   It*'s costing* us . . .

   These sentences tell us about the TEMPORARY situation.

B. **Present Simple** — to talk about how the system *works in theory or as specified*
   e.g. We *send* files
   We *start* transmitting
   It *takes* 5 minutes
   The files *are stored*

   These sentences tell us about the CHARACTERISTIC situation. The sentences are introduced with expressions like:

   in theory, ...
   each time, ...
   say (for example), ...
   when/if . . . .

   Note the 's' in the third person singular: It takes . . .

3 Controlled Practice

A. Put the verbs in the right tense:
   1. When we _ _ _ _ (send) a file, it _ _ _ _ (cost) us £1.00 a minute.
   2. At the moment, it _ _ _ _ _ _ _ _ (cost) us too much.
   3. I can't send the file now. The system _ _ _ _ _ _ _ _ (not work).
   4. In principle, the client _ _ _ _ (access) the file when he _ _ _ _ (want).
   5. Can you come and help? I _ _ _ _ _ _ _ _ (have) a lot of problems with this program.
   6. For example, we _ _ _ _ (transmit) a file in the morning. The client _ _ _ _ (get) the file out of the mailbox in the afternoon.
   7. Currently we _ _ _ _ _ _ _ _ (spend) too much time transmitting files.
   8. Breakdowns on the line _ _ _ _ _ _ _ _ (not happen) very often.
   9. We _ _ _ _ _ _ _ _ (lose) a lot of data because of this breakdown.

B. Correct these sentences, if necessary.
   1. The Dialcom Service is operating from a large computer centre in London.
   2. When you subscribe to the service, all your terminals can be connected to the mailbox.
   3. You are controlling the password which is giving the user access to his own private electronic office.
   4. The Data Network is providing access at many centres around the UK.
   5. At the moment, Dialcom rapidly expands its international network.

4 Transfer

PAIRWORK

Student B: Turn to Key Section

Student A: You represent a modem manufacturer. Below are the specifications for one of your products:

MODEM TYPE:	V22
SPEED OF OPERATION:	1200 bits per second
MODE OF OPERATION:	Answer mode or Call mode
	Full duplex
CONNECTION to COMPUTER:	RS232C cable
CONNECTION to LINE:	Telephone socket
AUTOMATIC or MANUAL DIAL-UP	

One of your customers is having problems with his modem. Ask him how he is using it at present: e.g. What speed are you using?
Then tell them what the specification is:
e.g. It works at 1200 bits per second
It is designed to work . . .

5 Word Check

to link up — to connect
to transmit — to send
to corrupt — to change, add or delete (data)
to subscribe — to pay a monthly/yearly amount to be a member
a subscription
configuration — system design/organisation
software package — software program(s) produced for general sale
call charge — fixed amount of money for a telephone call (depends on time and distance)
the weak link — the part of a system which will break down, cause problems

UNIT 16 Information Retrieval

(tenses – present and future)

Introduction

This unit deals with *retrieving* (getting back) information from a microfilm library. To create a microfilm library, documents are photographed and then *miniaturized* (made very small). A *microfilm reader* is a device for looking at microfilm records. The operator can search for the document quickly and then look at a magnified picture of the microfilm.

1 Listening 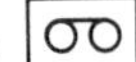

You are going to hear a presentation about a new information retrieval system. The system is based on microfilm. As you listen, label the system diagram below:

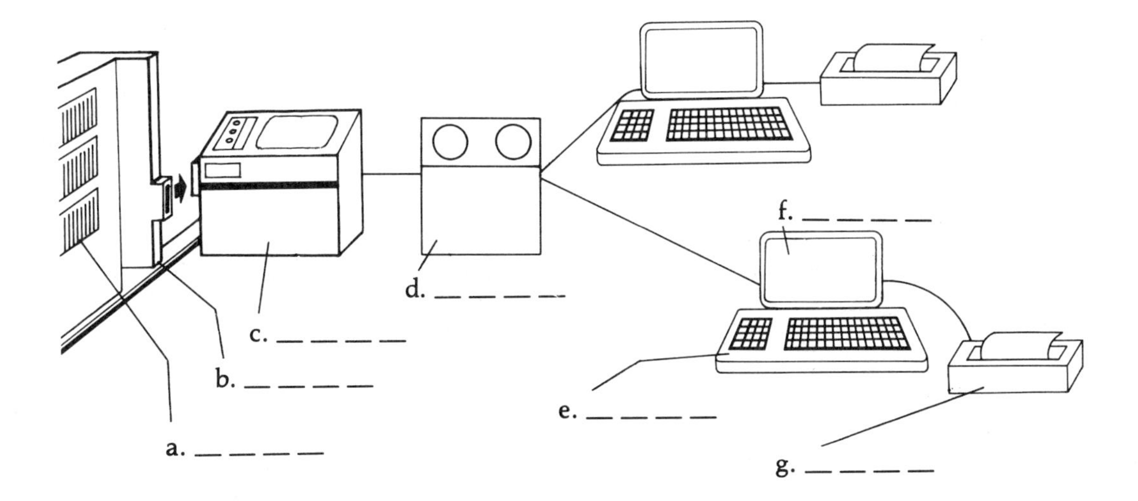

2 Presentation

The presenter talked about the information system from 4 points of view:

A. Present State
He used the *Present Simple*:
e.g. The clerk *inserts* the microfilm index.
He *searches* for the document.

B. Future State
He used *will*:
e.g. It'*ll* consist of a terminal . . . (Note contraction: *will* becomes *'ll*)
The image *will* be scanned.

C. Current/Present Work
He used the *Present Continuous*:
e.g. We'*re doing* a trial.

D. Future Plans
He used *going to, plan to, expect to*
e.g. We'*re going to instal* work stations.
We *plan to instal* it . . . .
We *expect it to come on line* . . . .

3 Controlled Practice

Complete these sentences. If necessary, listen again to the tape:

1. The operator __ __ __ __ down to the microfilm library.
2. He __ __ __ __ the microfilm index into the microfilm reader.
3. He __ __ __ __ the relevant microfilm.
4. He __ __ __ __ in the reader for the document.
5. If he __ __ __ __, he can take a thermal copy of the document.

6. The operator __ __ __ __ __ __ __ __ up the index on his terminal.
7. The index __ __ __ __ __ __ __ __ on disc in a central computer.
8. The autoloader __ __ __ __ __ __ __ __ the right microfilm from the library.
9. The image __ __ __ __ __ __ __ __ electronically __ __ __ __ and then transmitted to the terminal.
10. The operator __ __ __ __ __ __ __ __ the document on his terminal screen.

11. At the moment we __ __ __ __ __ __ __ __ a trial on the autoloader.
12. We __ __ __ __ __ __ __ __ on some new software.

13. We __ __ __ __ __ __ __ __ __ __ __ __ __ __ __ __ the work stations by the end of the year.
14. We __ __ __ __ __ __ __ __ __ __ __ __ the autoloader at the beginning of next year.
15. We __ __ __ __ the whole system __ __ __ __ __ __ __ __ __ __ __ __ __ __ __ __ in the spring next year.

4 Transfer

PAIRWORK

Student B: Turn to Key Section

Student A: Below is a diagrammatic representation of information flow between the Head Office and Sales Branches in your company. Describe this to Student B as the 'present state' and then tell him/her about current work:

1 Training: word processing for secretaries.
2 Research: cost and operation of electronic mail.

Student B will then tell you about the 'future state' and project timing.

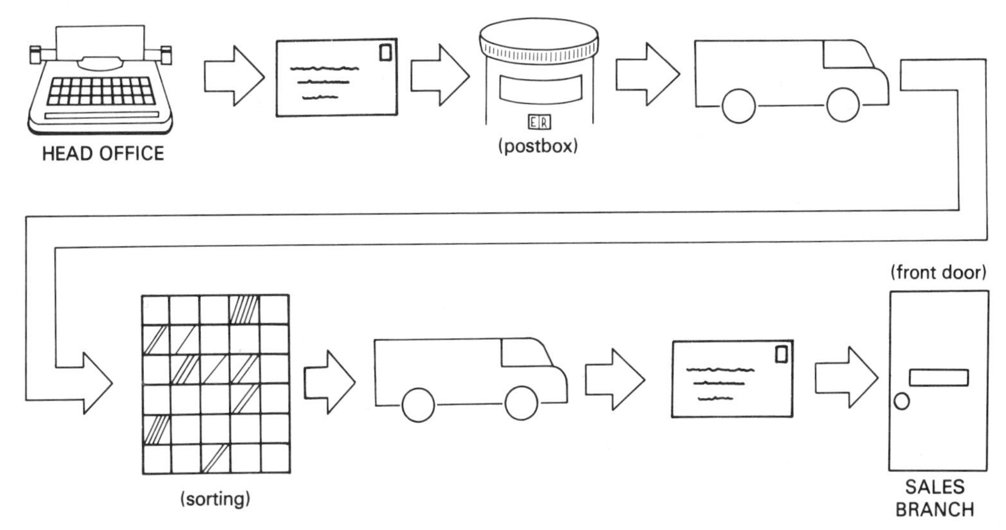

5 Word Check

PROCESS VERBS
to insert — to put in
to select — to choose
to search — to look for
to install — to put equipment in place
to call up — to ask the system to show some information
to scan — to detect and register text using a photoelectric device
to transmit — to send
to view — to look at
to do a trial — to test equipment for a longer time (several weeks/months)
to come on line — to start operating fully

EQUIPMENT/OUTPUT
a thermal copy — a heat-printed copy
a hard copy — a dry printed copy
(on) disk — magnetic/floppy disk
an autoloader — a robotic device which takes the microfilm cassette from the library and puts it in the reader

UNIT 17 International Aviation Standards

(nouns, adjectives and adverbs)

Introduction

This unit deals with international aviation standards and mentions 6 objectives that the industry should work towards.

1 Listening 

You are going to hear a talk given by a senior aviation official. As you listen, complete the table below. Write the areas appropriate to the 6 goals in the right-hand column. The first one has been done for you.

Goals	*Areas*
efficiency	Air Traffic Control Service
safety	
speed	
precision	
reliability	
comfort	

2 Presentation

Here is some of the language the speaker used.

Noun	*Adjective*	*Adverb*
efficiency	efficient	efficiently
safety	safe	safely
speed	quick/fast	quickly/fast
precision	precise	precisely
reliability	reliable	reliably
comfort	comfortable	comfortably

We use an adjective to give more information about a noun, e.g. A reliable service.
We also use an adjective after the verb 'to be', e.g. The service is reliable.
We use an adverb to give more information about a verb, e.g. The service must operate reliably.
Adverbs normally end with 'ly', e.g. reliably.

3 Controlled Practice

Now complete the following sentences by putting the word in brackets into the correct form — noun, adjective or adverb.

Life on an offshore oil drilling platform used to be (uncomfortably) ___1___ and (danger) ___2___.

But in recent years these conditions have improved (substantial) ___3___. For oil rig employees (reliability) ___4___ information about conditions affecting their work and the weather is needed (quick) ___5___. Of course much information can now be provided (efficiency) ___6___ and (precision) ___7___ by the advanced instruments currently in use. So, although the (safe) ___8___ of the workers cannot be guaranteed and accidents (occasion) ___9___ happen, working and living conditions are (usual) ___10___ quite (acceptably) ___11___. If a serious accident occurs, it is (essentially) ___12___ to evacuate the rig as (quick) ___13___ as (possibility) ___14___.

As for living conditions, it is now (normally) ___15___ practice to accommodate the rig workers in reasonable (comfortably) ___16___, although studies show that noise levels from equipment can sometimes be quite (height) ___17___. However, these disadvantages are offset by the fact that this equipment helps the workers to carry out their jobs with (efficient) ___18___ and (precise) ___19___. And at the end of their working shift they are transported (quick) ___20___ to the mainland for leave.

4 Transfer

GROUP WORK

Discuss the role and importance of the following factors in relation to a technical area or product which you know:

- efficiency
- safety
- reliability
- precision
- speed
- comfort

5 Word Check

CHANGE
development — change (for the better)
improvement (also to improve) — better level, condition, etc.
innovation — new idea or product
advance — change (for the better)

STANDARDS
efficiency — ability to work well
safety — freedom from danger
speed — ability to work quickly
precision — ability to work exactly
reliability — ability to work without breaking down
comfort — freedom from body pain

OTHER WORDS
procedures — actions to do something properly
to grumble — to complain
to delay — to make late
to shrink — to make small
to respond — to act
caring — friendly and sympathetic
courteous — polite

UNIT 18 Scuba Diving

(conditional I and II)

Introduction

This unit deals with a computer for scuba divers which gives information about time, depth, and *air supply*; it also warns divers about critical conditions. Some of the information is *visual* and some is *audible*.

1 Listening

As you listen, tick the appropriate column for each action or situation, depending on whether the computer provides visual information or an audible warning.

Action/Situation	*Visual information*	*Audible warning*
1. Go to the surface now.		
2. You are 50 metres below the surface.		
3. Battery level – OK.		
4. Seawater dive.		
5. You started your descent 10 minutes ago.		
6. Look at your AIR TANK CONTENTS display.		
7. Slow down your ascent.		
8. The amount of air in your tank is 144 bars.		
9. You started to decompress 4.28 minutes ago.		

2 Presentation

Here is some of the language you have just heard. The speaker used two kinds of conditional sentences.

A condition	result
If a diver *needs* to spend 5 minutes at a depth of 5 metres for decompression,	he *will start* his stop-watch and . . .
If the batteries *are* low	the instrument *will* not *function* at all.
If a diver *wants* an accurate depth reading,	he *must set* the appropriate water type.
present tense	future with *will* or modal, e.g. *can, must, should,* etc

B condition	result
If he *began* his ascent . . .	he *would be* within safe limits.
If he *went* up too quickly,	he *would see* a warning light.
If a diver *wanted* to know . . .	he *could see* this from the contents display.
past tense	conditional tense with 'would' or modal e.g. *could*

The difference between A and B is based on the speaker's attitude. Both A and B express a relationship between a condition and a result. In A the speaker expresses the condition and result as real. In B the speaker expresses the condition and the result as possible or hypothetical. A is a real condition (R); B is a hypothetical condition (H).

3 Controlled Practice

Now look at the information in the table below, and make sentences using the conditional constructions. If there is an R (real), make a sentence based on model A above. If there is an H (hypothetical), make a sentence based on model B above. The first one has been done for you.

		condition	result
1	R	You are not under water	the air pressure varies according to your altitude
2	H	You are under water	the pressure increases
3	H	The pressure increases	you use your air more quickly
4	R	You are under water	you need to know exactly the air time left
5	H	You stay under water for a long time	you need to spend time in decompression
6	R	You don't spend time in decompression	you may suffer from the bends
7	R	You want to avoid decompression	you must start your ascent within one minute
8	H	You don't start your ascent then	you have to stay at 5 metres for 10 minutes

9	R	You look at your diver's watch	you see your present depth in metres
10	H	You see the 'ASCENT TOO FAST' display	what do you do?

1. If you are not under water, the air pressure will vary according to your altitude.
2. ___________________________________
3. ___________________________________
4. ___________________________________
5. ___________________________________
6. ___________________________________
7. ___________________________________
8. ___________________________________
9. ___________________________________
10. ___________________________________

4 Transfer

GROUP WORK

The device described on the tape is designed for hobby users (up to a maximum depth of 100 metres) – not for commercial divers. Discuss the real and possible/hypothetical uses for such an instrument. Consider the following:

- applications (looking at fish, wrecks, underwater archaeology, etc.)
- sea/weather conditions
- limitations (hobby not commercial, no deeper than 100 metres, etc.)

Use the 2 types of conditional sentences,
e.g. A diver can use this device if he wants to . . . .
A diver wouldn't use this device if he wanted to go lower than 100 metres.

5 Word Check

INSTRUMENTS AND EQUIPMENT

gauge – instrument for measuring something
display – part of a device which shows a piece of information
tank – container for air
signal – sign indicating information (in this case of danger)

OTHER WORDS
to descend (descent) – to go down
to ascend (ascent) – to go up
to decompress (decompression) – to reduce the pressure on someone
to elapse – to pass (of time)
to transmit – to send
to flash – to shine a light on and off for a very short time

UNIT 19 Control Systems

(adjectives and adverbs)

Introduction

This unit deals with crane control systems. All the cranes are the overhead type. However, some can be controlled from the ground (*floor control*), others from a cab (*cab control*).

1 Listening

A technical sales representative is talking to a customer about cranes. He describes the different types of control systems. As you listen, label the diagrams with right descriptions from the list below:

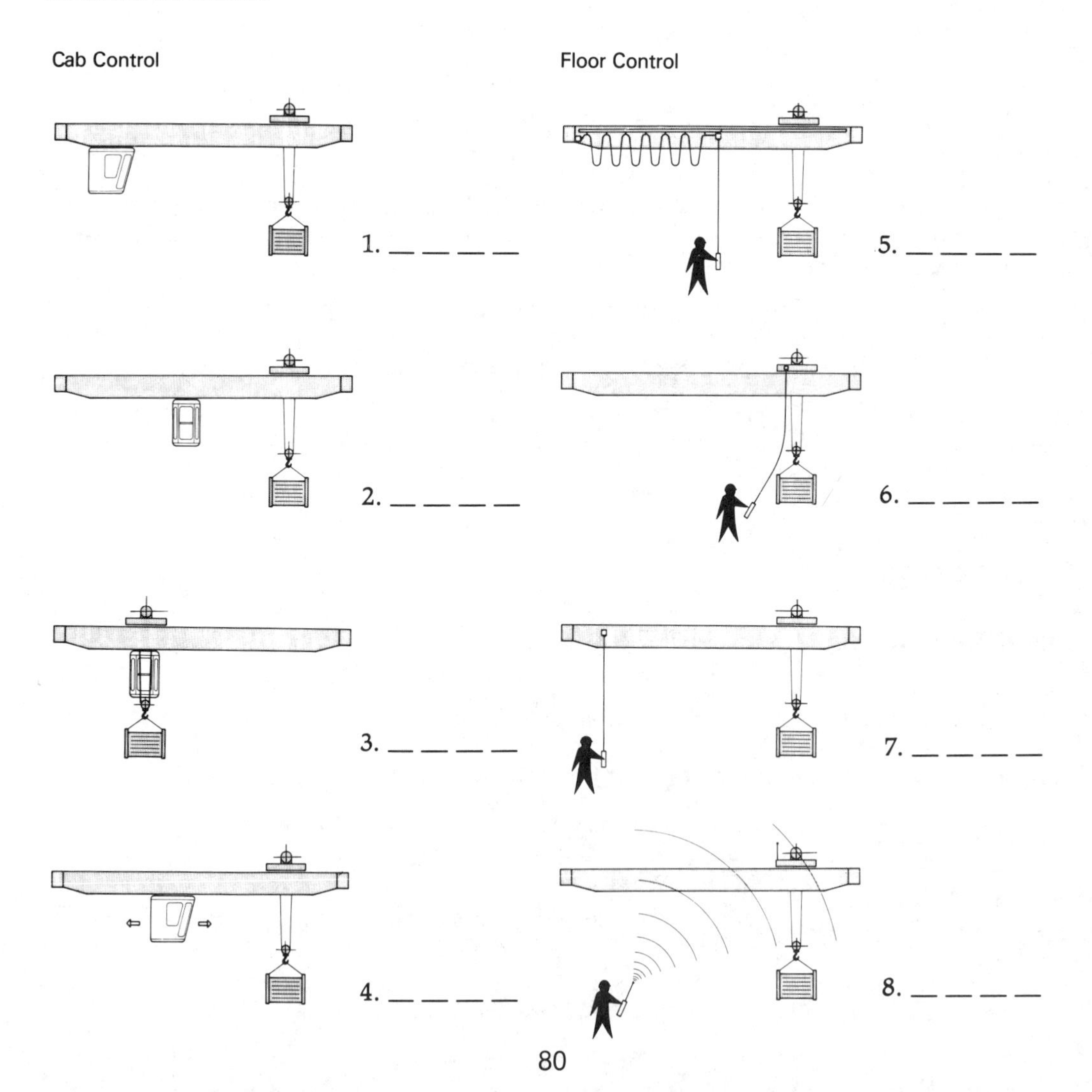

Descriptions

a. independently mobile cab
b. mobile control pendent
c. remote control
d. laterally mounted cab
e. crab suspended control pendent
f. crab mounted cab
g. centrally mounted cab
h. laterally suspended control pendent

---

2 Presentation

Look at how these complex descriptions are built up:

	adverb	*adjective*	*classifiers*	*noun*
a	______	______	__________	device
a	______	small	__________	device
a	______	small	______ control	device
a	______	small	remote control	device
an	extremely	small	remote control	device

NOTES: 1. the *adjective* and *classifier* describe the *noun*
2. the *adverb* describes the *adjective*

3 Controlled Practice

A. Put these in the right order:

1. hot / conditions / operating / extremely
2. mounted / a / cab / centrally
3. a / control / manually / guided / pendent
4. console / a / control / stationary
5. visibility / improved / substantially

B. Make noun phrases as alternatives
e.g. A crane which is supervised by a man
⇒ A manually supervised crane.

1. Conditions of operation which are extremely unpleasant.

   _

2. An application in the industrial sector which involves heavy loads.

   _

3. An engineer who has good qualifications in electronics.

   _

4. A process in the chemical industry which is controlled by a robot.

   _

5. A device which was cheap to manufacture.

_ _

6. Software which has been specially developed for the customer.

_ _

4 Transfer

Student B: Turn to Key Section

Student A: Find out more about the following by asking Student B questions. Together build up a noun phrase to describe the object.

e.g. A car	Question: What type?	Answer: Sports
	Q: What size?	A: 2000 cc
	Q: Where is it made?	A: Japan

⇒ *A 2000 cc Japanese sports car*

1. *A machine* — Questions: What type?
   What size?
   How does it operate?

   ⇒ _

2. *A drill* — Questions: What type?
   How does it work?

   ⇒ _

3. *A computer* — Questions: Type?
   Size?
   Origin?

   ⇒ _

5 Word Check

operating conditions — atmosphere/climate in which work is done
visibility — condition for seeing things at a distance
a crab — lifting equipment (chains, etc.)
supervision — watching and directing
a load — anything carried or supported (in this case by a crane)
stationary — not moving
a console — a terminal with operating switches/buttons
to mount — to fix on something
to vary — to change
to guide — to help find the right way/position

UNIT 20 Project Planning

(prepositions of time)

Introduction

This unit deals with the timing of a project. The project leader is responsible for different activities including *installation* of equipment, *testing* of the equipment and *training* of the employees.

1 Listening 

You are going to listen to a phone call. Jim, manager of overseas operations of an engineering firm, calls Chris, project leader in charge of setting up a new plant in Saudi Arabia. They discuss the schedule for the project. As you listen, complete the key for the 'Planner' below:

May		June				July					August				September		
21	28	3	10	17	24	1	8	15	22	29	4	11	18	25	1	8	15

Today: Thursday 6th June

Key:

a. ___________

b. ___________

* f. ___________

□ g. ___________

○ h. ___________

Training Courses

c. ___________

d. ___________

e. ___________

2 Presentation

During the telephone call, some of the following time expressions were used:

TIME RELATING TO NOW

DAYS

2 days ago ← yesterday ← **today** → tomorrow → next Monday → a week from next Monday

WEEKS/MONTHS

2 weeks ago ← last week ← **this week** → next week → in 2 weeks' time (the week after next)

NOTE: With *this* . . . , *next* . . . , *last* . . . , NO preposition is used.

DAYS/DATES/MONTHS, etc.

on Monday (days of the week)
1st August (dates)

at 12 o'clock (times)
the beginning of . . .
the end of . . .
the weekend (NOTE: American English: *on* the weekend)
night

in July (months)
the middle of . . .
1987 (years)
the morning/afternoon/evening
autumn/winter/spring/summer

by Monday (= at the latest)

3 Controlled Practice

A. Use the planner to complete these sentences:

1. Installation work finished __ __ __ __ week.
2. Testing will begin __ __ __ __ Monday __ __ __ __ week.
3. Testing will finish __ __ __ __ __ __ __ __ __ __ __ __ time.
4. The first operators training course began __ __ __ __ Wednesday __ __ __ __ week.
5. It'll finish __ __ __ __ week.
6. The first maintenance course begins a week __ __ __ __.
7. The supervisors' course finishes __ __ __ __ __ __ __ __ __ __ __ __ __ __ __ __ July.
8. The plant will start up __ __ __ __ 25th August.
9. We plan to reach full capacity __ __ __ __ September 8th (at the latest).
10. The plant will be officially opened __ __ __ __ September 15th.

B. Write out the following telex in full:

ATTN: Chris Kerridge, Project Leader
I cnfm F. Hyman arrives Riyadh Sat. 1200. He expects start training 17 June. Pls meet him airport. 20 Operators manuals sent ysdy. Should arr. beg. nxt week.
Rgds
Jim Coleman

Attention: Chris Kerridge, Project Leader

I confirm ____

____

____

4 Transfer

PAIRWORK

Student B: Turn to Key Section

Student A: Fill in the planner below by asking Student B questions. The planner should illustrate the project timing (including consultancy, construction and installation)

October					November					December			
3	10	17		24	1	8	15	22	29	6	13	20	27

Today, Wednesday 19th October

5 Word Check

to schedule — to plan activities in time
to be on schedule — to put plan into action at the right time
to kick off — to start
to be due to — to be planned/scheduled to ...
to start up — to start (a process, factory, etc.)
to build up capacity — to increase production volume gradually
to reach full capacity — to arrive at the moment when the factory is producing at full volume
a planner — see chart in section 1
a diary — daily record of appointments, meetings, etc.

UNIT 21 Speeding up Air Traffic

(prepositions of location and movement)

Introduction

This unit deals with a computer system developed for a British airport to speed up air traffic. The system handles flight information, passenger check-in and point-of-sale terminals in the duty-free shop.

1 Listening

As you listen, complete the sentences in the table below. In each sentence you need to write the appropriate preposition. The first one has been done for you.

Facilities (handled by the computer system)	*Where*	
50 Visual Display Boards have been installed	in	the main concourse
	___	the arrival and departure lounges
	___	arrival and departure gates
	___	the baggage collection areas
Flight information is also displayed	___	the airport's closed-circuit TV system
A character is formed by light shining	___	the required holes
We have 14 check-in points	___	the whole of one side of the main concourse
A passenger can go	___	any check-in point
Our duty-free shop is	___	the departure lounge
Each duty-free item has the price coded	___	the price tag
One copy of the receipt goes	___	the customer
After buying his duty-free a passenger can go	___	the shop and
	___	the lounge

2 Presentation

In the listening passage you heard various prepositions such as *in*, *on* or *at*. These were used as parts of expressions of *location* or of *movement*.

The picture below of the inside of an airport terminal illustrates some of these expressions of location (L) and movement (M). As you will see, the preposition we use usually depends on our 'view' of the places we are talking about.

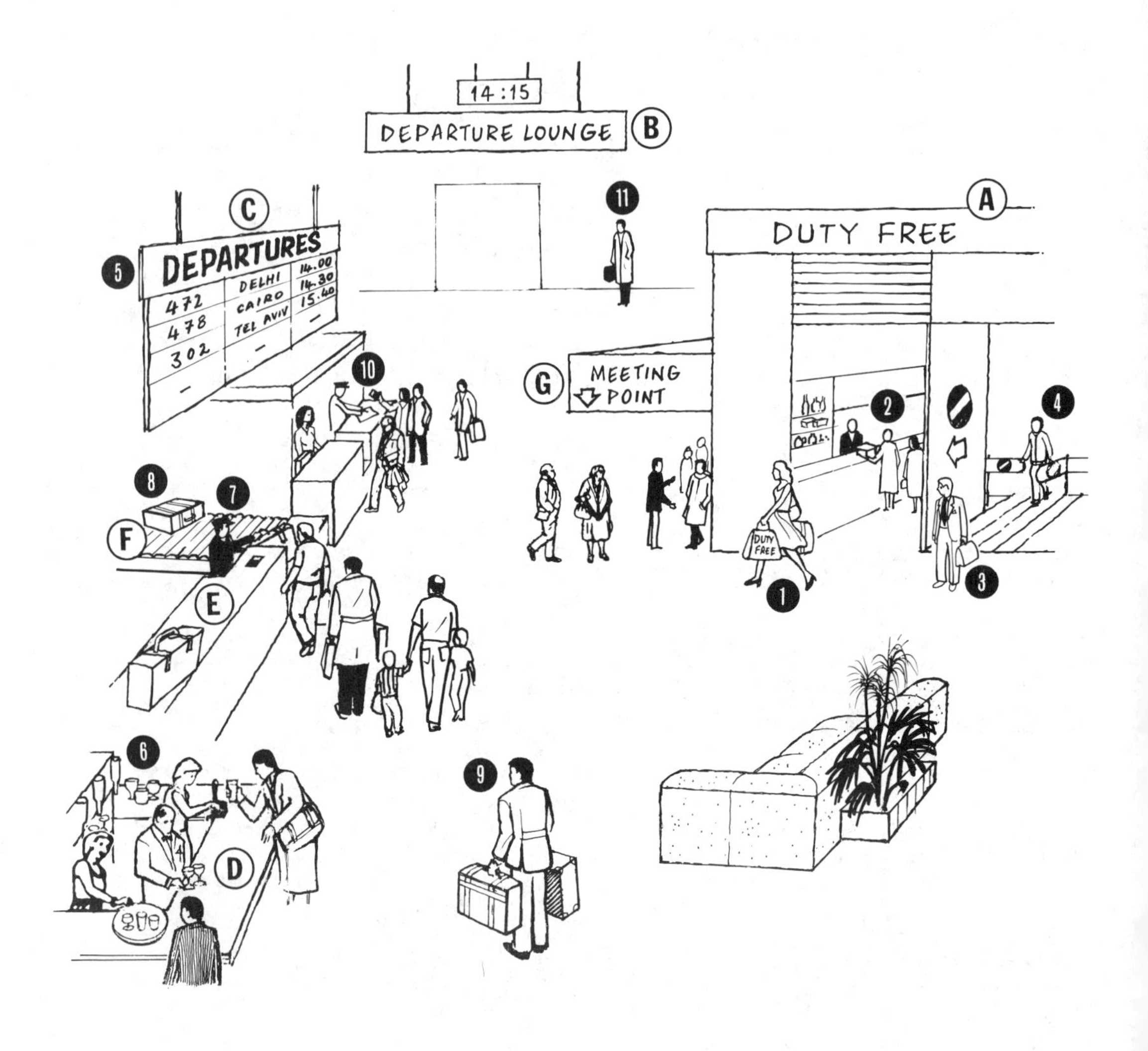

A. IN, INSIDE, OUT OF, OUTSIDE, THROUGH

Look at the duty-free shop (A) and the departure lounge (B) in the picture. We can view these as places with the three dimensions of height, length and depth. Now look at passengers (1), (2), (3) and (4), and study the sentences below.

A passenger (1) is coming OUT OF the duty-free shop and (going) INTO the departure lounge. (M)

A passenger (2) is IN (INSIDE) the shop buying cigarettes. (L)

A passenger (3) is waiting OUTSIDE the shop for his companion. (L)

A passenger (4) has not bought anything. He is walking THROUGH the check-out. (M)

B. ON (ONTO), OFF, ACROSS, OVER

Now look at the electronic display board (C) and the counter (D). We can view these as places with the two dimensions of length and width (but no depth and no height). Now look at the plane departure announcements (5) and the drinks (6).

The departure time of Flight 478 has just been indicated ON the display board. (L)
A passenger is taking a glass of beer OFF the counter. (M)
The barman is putting some glasses of wine ON(TO) the counter. (M)
The departure time of the Delhi flight will very soon be taken OFF the display board, because the flight has already departed. (M)
A waitress is pushing a tray of drinks ACROSS (OVER) the counter to a customer. (M)

C. OFF, ON, ALONG, ACROSS, OVER

Look at the check-in desks (E) and the conveyor-belt (F). We can view these as places with only one dimension of length (and no width or depth). Now look at the suitcases (7) and (8) and the check-in clerks in the picture.

One check-in clerk is taking a suitcase (7) OFF the scales and putting it ON(TO) the conveyor belt. (M)
A suitcase (8) is ON the conveyor belt and is travelling ALONG it to the loading bay. (L and M)
A passenger (9) is looking ALONG the line of check-in desks for a vacant one. (M)
A passenger (10) has just handed her passport and ticket ACROSS (OVER) the desk to the clerk. (M)

D. AT, FROM, TO

Look at the meeting point (G). We can view this as a place without any specific dimensions. The examples below show how many of the places in the airport can be viewed in the same way.

Some people are waiting AT the meeting point. (L)
Most passengers in the picture are waiting in queues AT the check-in desks. (L)
A passenger (11) is carrying a bag of duty-free shopping FROM the arrivals lounge TO the meeting point. (M)

3 Controlled Practice

A. Now write in the box beside each preposition (L) for Location or (M) for Movement.

1. When you arrive inside ( ) the airport building, go to ( ) the check-in desk.
2. Make sure you have taken your ticket, passport and money out of ( ) your baggage, before you put it on ( ) the scales.
3. The clerk at ( ) the desk will check your ticket.
4. She will tie a baggage label onto ( ) the handles of your suitcases.
5. Then you can either wait in ( ) the main concourse or go through ( ) passport control to ( ) the departure lounge.
6. At ( ) most airports you will find a duty-free shop and a restaurant in ( ) the departure lounge.
7. Do not take the seal off ( ) your duty-free purchases before you get on ( ) the plane.
8. Once inside ( ) the plane, stow your hand baggage safely: do not put any heavy bottles into ( ) the overhead lockers.

B. Now complete the following messages between ground control and the captain of an aircraft.

1. This is Flight 307 __ __ __ __ Madrid.
2. Hold __ __ __ __ Gate Two.
3. Taxi __ __ __ __ position and hold.
4. Wait for further instructions __ __ __ __ the tower.
5. Proceed to holding point. Clear for take-off. Start taxiing __ __ __ __ the runway.
6. Contact departure control __ __ __ __ frequency 128.0.
7. Climb __ __ __ __ flight level 120.
8. You are now clear to move __ __ __ __ British controlled airspace into French airspace.
9. You are about to pass __ __ __ __ moderate turbulence.
10. We are __ __ __ __ __ __ __ __ turbulence now. Clear conditions ahead.

4 Transfer

GROUP WORK

Discuss where you would site a new airport, railway station or motorway/major road for your city or region. Consider the following:

- location
- access
- public transport
- facilities, e.g. car parks, check-in, baggage handling, restaurants, duty-free shopping, etc.

5 Word Check

concourse — large open area, e.g. in an airport, railway station, etc.
to repeat — to give information again
closed-circuit TV system — system which sends information to a limited number of TVs
to generate — to produce
standard — normal (the opposite is non-standard)
message — piece of information
to be composed of — to consist of
matrix — arrangement of holes in rows from which a letter or number can be produced
to shine — to direct or send light
flexible — can change for new needs or conditions
reliable — can be trusted, doesn't break down too often
flap — board which shows arrival and departure information
to represent — to show
integrated — combined, not individual
to code — to translate into a special system, e.g. numbers, bars, etc. so that it can be read (electronically)
(price) tag — small piece of paper or material with information, e.g. of price
beam — line of light

UNIT 22 Electronic Assembly

(comparison of adjectives)

Introduction

This unit deals with a new method of fixing electronic components to a circuit board. In the *conventional* (usual) method, components are fixed through holes in the board. In the new method (*surface mounting*) the components are fixed directly onto the board without any holes.

1 Listening 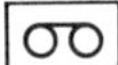

A consultant engineer is answering questions about a new method of assembling electronic components on a circuit board – 'surface mounting'. As you listen, tick ( ) if the feature of the surface mounted board is an *advantage*; put a cross (X) if the feature of the surface mounted board is a *disadvantage*.

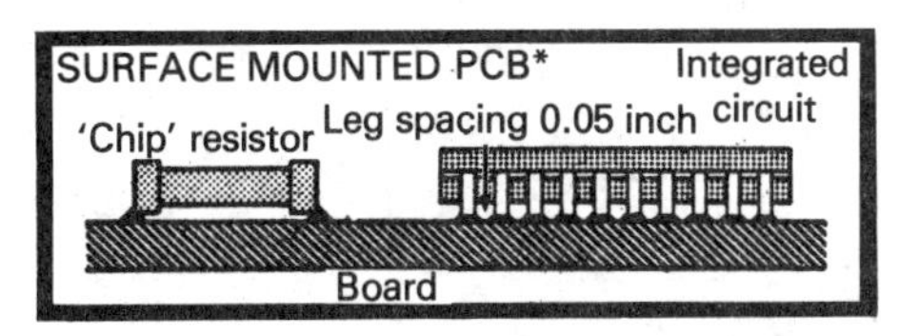

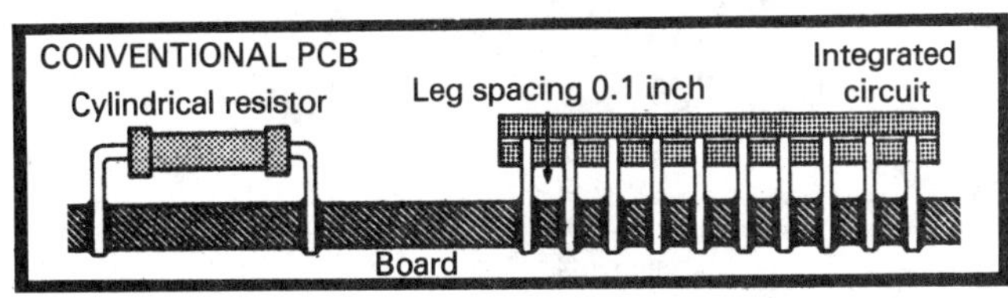

Features	*Surface-mounted versus conventional assembly*
Size	
Weight	
Cost of production	
Speed of assembly	
Cost of assembly	
Reliability	
Accuracy of placement	
Autotesting	

* Printed circuit board

2 Presentation

The consultant engineer *compared* the two processes from many points of view. In making comparisons, he followed these rules:

A. *1 syllable adjectives*: add *'er'*
e.g. SMALL — SM components are much *smaller* than conventional components

B. *2 syllable adjectives ending in: 'ow', 'le' and 'y'* — add: *'er'*
e.g. GENTLE — The SM assembly process is much *gentler*
EASY — The SM assembly process is *easier* (NOTE change *'y'* to *'i'*)

Other 2 syllable and 3/4/5 syllable adjectives: use *'more'*
e.g. ROBUST — Old components had to be *more robust*
EXPENSIVE — New components are *more expensive* to produce

3 Controlled Practice

A. Complete the table

Feature	*Adjective*	*Opposite adjective*
Weight	______	light
Size	______	______
Width	______	narrow
Ease	______	______
______	fast/quick	______
______	______	unreliable
Accuracy	______	______
______	______	inefficient
Cost	______	expensive
Strength	______	______

B. Complete the sentences:

1. *Size*: Conventional components are _ _ _ _ than SM components.
2. *Weight*: SM components are considerably _ _ _ _ than conventional components.
3. *Cost of production*: Conventional components are a little _ _ _ _ to produce.
4. *Speed of assembly*: SM assembly is much _ _ _ _ than conventional assembly.
5. *Cost of assembly*: Conventional assembly is _ _ _ _ _ _ _ _ than SM assembly.
6. *Reliability*: SM components are considerably _ _ _ _ _ _ _ _.
7. *Accuracy of placement*: The SM assembly process is _ _ _ _ _ _ _ _ than conventional.
8. *Autotesting*: At the moment, autotesting is _ _ _ _ with conventional components.
9. *Width of pcb*: SM pcbs are _ _ _ _ than conventional boards.
10. *Strength of components*: Conventional components must be _ _ _ _ than SM components because the assembly process is not so gentle.

4 Transfer

PAIRWORK

Student A: Below are some specifications for a particular SM board. Student B has the specifications for a conventional board. Ask him/her for these specifications.

Student B: Look in the Key Section for your specifications. Then ask Student A for his/her specs.

	SM Board	*Conventional board*
Size	l: 10 in w: 5 in	
Thickness	0.8 cm	
Weight	480 gms	
MTBF (mean time between failure)	1,400 hours	
Accuracy of component placement:	± 0.005 in	
Testing ratio (autotest: manual test)	80 : 20	
Assembly method autoinsertion: machine insertion: manual insertion:	 50% 40% 10%	

After you have completed the table, make comparisons between the SM and conventional boards.

5 Word Check

PROCESS VERBS

to assemble — to fix pre-manufactured parts to the main structure
to drill (holes) — to make a hole in a surface
to push through — to insert in or through a hole
to solder — to join using a metal alloy (Sn + Pb)
to autoinsert — to push components through the holes automatically
to vibrate — to move from side to side or up and down; to shake continuously
to autotest — to test components automatically
to tolerate — to be able to work under stress (e.g. vibration)

QUALITIES

robust — strong, well-built
gentle (process) — not needing to be strong
reliable — can be trusted, doesn't break down often

UNIT 23 Energy Sources

(question formation)

Introduction

This unit deals with various energy sources and generation processes. It compares the most efficient nuclear methods with *conventional fossil fuel* methods.

1 Listening 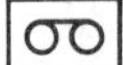

In this dialogue an energy specialist is answering questions about different sources of energy. He compares how long 10 kilograms of different kinds of fuel will last.
As you listen, write down the length of time for each type of energy source/process.

Energy generation per 10 kgs in a 2 million kilowatt power station

Fuel/Process	*Running time per 10 kgs*
Nuclear Power	
Hydrogen Fusion Reactor	
Fast Reactor	
Natural Uranium	
Oil	
Coal	

2 Presentation

QUESTION FORMATION FLOWCHART

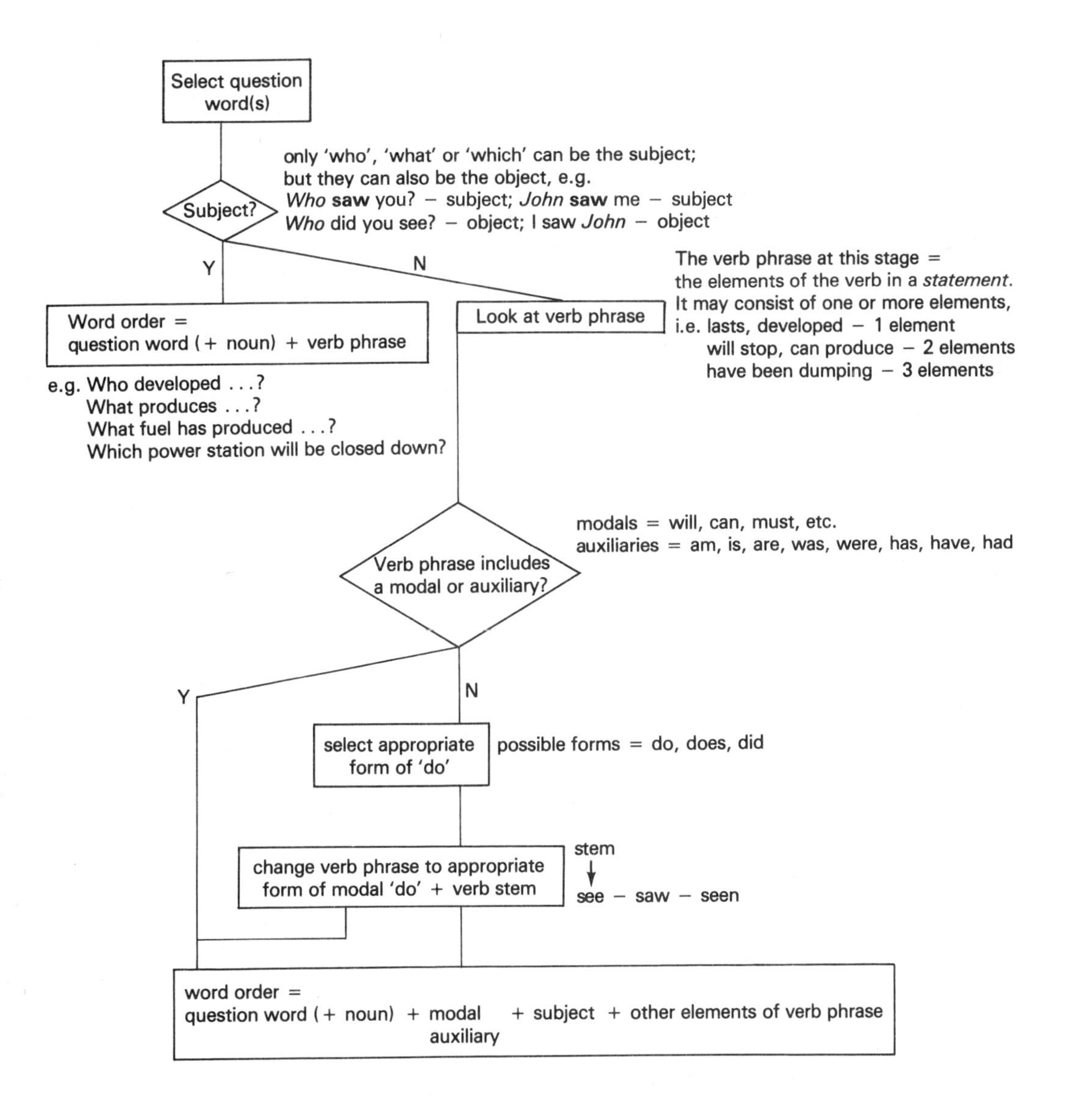

e.g. How long	does	it	last?
When	will	it	stop?
Why	did	we	develop . . . ?
Where	have	we	been dumping . . . ?
How much fuel	does	it	use?
How many kilowatts	does	it	generate?
Who	can	we	sell it to?
What	do	we	know about . . . ?
How	have	we	generated . . . ?

3 Controlled Practice

Now use the Question Formation Flowchart to make questions using the information in the table below. The first one has been done for you.

Question word/phrase	*Subject*	*Verb Phrase*	*Other sentence elements*
How much energy	we	use	per day
How	we	generate	this energy
Why	nuclear power	is becoming	more important
When	we	started	to use nuclear power
What	we	do	with nuclear waste
How long	we	have been dumping	nuclear waste in the sea
Where else	we	can dump	nuclear waste
Which	country	produces	most nuclear energy
How many nuclear power stations	that country	has	
Where	that country	dumps	its nuclear waste
When	fossil fuels	will run out	
How long	we	can survive	without nuclear energy

1. How much energy do we use per day?
2. ______________________________
3. ______________________________
4. ______________________________
5. ______________________________
6. ______________________________
7. ______________________________
8. ______________________________
9. ______________________________
10. ______________________________
11. ______________________________
12. ______________________________

4 Transfer

PAIRWORK

Ask and answer questions about:

- the main types of fuel used to generate energy in your country
- the main types of process used to generate energy in your country
- the energy provision in your area
- the energy requirements in your area
- the future energy requirements in your area

5 Word Check

ENERGY SOURCES AND GENERATION
nuclear reactor — machine for production of atomic energy
hydrogen fusion reactor — type of process for production of atomic energy
fast reactor — type of process for production of atomic energy
to convert — to change

UNIT 24 Factory Automation

(tenses – past, present perfect, present and future)

Introduction

This unit is about the automation of a factory. The process in the factory is divided into 3 stages:

- the supply of materials to the assembly line (*supply area*)
- the assembly line
- the *packing* and *sorting* area (where the assembled products are put in boxes and classified ready for distribution).

1 Listening 

Two production engineers present the developments in automation of a medium-sized factory. In their presentation, they talk about three phases of automation. As you listen, classify the following steps as:

A. First phase automation
B. Second phase automation
C. Third phase – option 1
D. Third phase – option 2

The first one has been done for you.

Steps	*Phases*
1. Installation of automatic packing equipment	A
2. Reduction in packing workforce from 6 to 2	
3. Introduction of automatic sorting	
4. Automation of assembly line	
5. Reduction of assembly workforce from 25 to 15	
6. $ ½ million investment in automatic assembly equipment	
7. Total automation of supply area	
8. Automatic picking from stores	
9. Automatic conveyor feeder	
10. Automation of component transport to workstations	
11. Partial automation of supply area	
12. Automatic conveyor feeder	

2 Presentation

In talking about the different phases, the engineers referred to different time periods:

1. The original state (2 YEARS AGO) and first phase automation (LAST YEAR)
2. Second phase automation (THIS YEAR)
3. Present state in the supply area (PRESENT SITUATION)
4. Third phase automation (FUTURE SITUATION)

We can show the relationships in time like this:

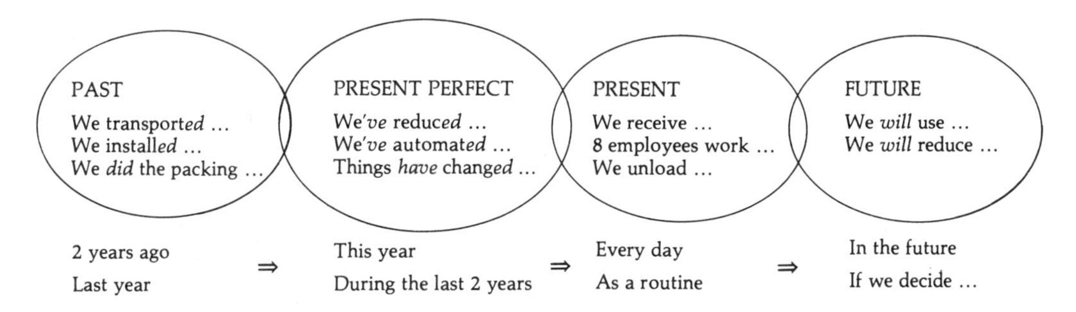

NOTES: 1. The events and the time are FINISHED when we use the PAST.
2. When we use the PRESENT PERFECT, the event may be finished but the time is UNFINISHED.
3. When the event *and* the time is UNFINISHED, we often use the PRESENT PERFECT CONTINUOUS, e.g. 'My team *has been* study*ing* the third phase'.

3 Controlled Practice

Complete the sentence by inserting a verb in the right tense.

A. Manpower

1. 2 years ago, 25 workers __ __ __ __ __ __ __ __ the assembly line.
2. In the packing department we __ __ __ __ the number of workers from 6 to 2.
3. This year we __ __ __ __ __ __ __ __ the assembly line workers from 25 to 15.
4. Altogether 8 employees normally __ __ __ __ in the supply area.
5. If we decide on total automation, we __ __ __ __ __ __ __ __ the workforce in the supply area from 8 to 2.

B. Automation

1. 2 years ago, we __ __ __ __ the packing and sorting by hand.
2. Last year we __ __ __ __ automatic packing equipment.
3. We also __ __ __ __ automatic sorting.
4. This year we __ __ __ __ __ __ __ __ the assembly line itself.
5. During the current year we __ __ __ __ __ __ __ __ $½ million in automatic assembly equipment.
6. We generally __ __ __ __ the motors from the trucks manually.
7. In the future, for the supply of components to the workstations, we __ __ __ __ __ __ __ __ microtrucks.

4 Transfer

PAIRWORK

Student B: Turn to Key Section

Student A: Tell Student B about the product development of the following. Try to answer these questions:

What did it look like?
How did it work?
How was it made?
How has it changed?
How has it improved?
What haven't they developed (yet)?
How is it made?
How will it change in the future?

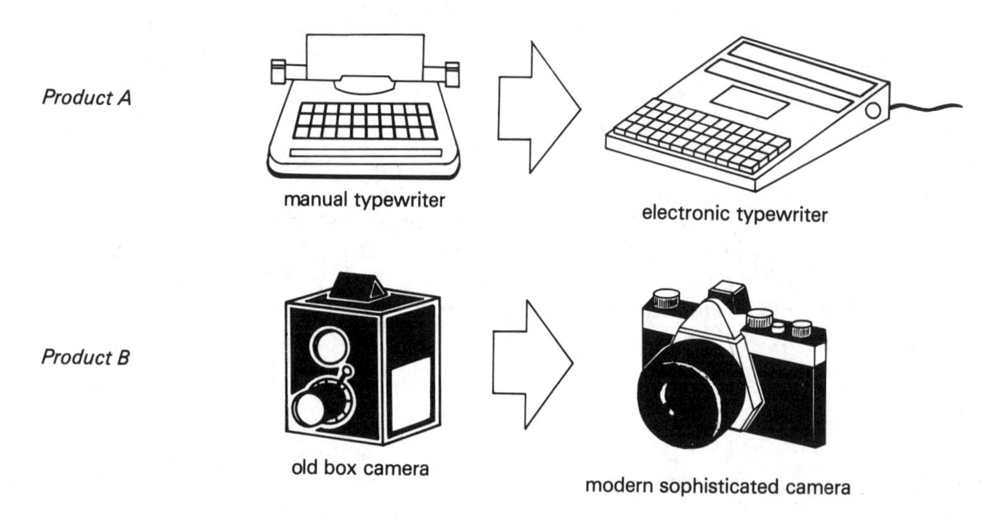

When you have finished, Student B will tell you about Products C and D.

5 Word Check

VERBS

to man — to operate a machine using an operator
  manual
  (by hand)
to automate — to change/convert a machine from manual to automatic operation
  automatic
  automation
to reduce — to decrease the number
to unload — to take products/goods off a truck
to store — to keep a quantity of goods ready for use
to pick — to choose and take a product/goods from a store

EQUIPMENT
a fork-lift truck –

a bar-code reader – a device using laser technology which identifies a product with a unique code. The code consists of a number of parallel lines.
a conveyor – moving belt for transport of products, etc.
a stock room – place where goods are stored
a tray – a small flat container used for carrying components
a feeder – a machine which introduces/enters a component onto a conveyor or into a process machine
a microtruck –

OTHER WORDS
a supplier – a company which sells raw materials, parts and components to a manufacturer

KEY SECTION Units 1–24

This section contains:

i tapescripts and keys to the Listening exercises
ii answers to the Controlled Practice exercises
iii information for the Transfer section where required

UNIT 1 A Bridge or a Tunnel?

1 Listening [cassette]

TAPESCRIPT

Good morning, ladies and gentlemen. We have received 5 plans for the 'Brunnel Project' from both local and foreign companies. As you will see from your diagrams the companies are split between 3 types of construction — a bridge link, a bridge/tunnel link, and a tunnel link.

This morning I would like to describe each of the plans briefly. Then we can have a fuller discussion about each proposal. You will then have time to consider each plan in detail, and I suggest we meet for further discussions 2 weeks from now. For reasons of confidentiality, I will not give the names of the companies. Instead I will simply call the plans number 1, number 2, number 3, number 4, and number 5.

To begin. As you can see from your diagrams, plan number 1 is for a construction 36 kilometres long at a height of 65 metres above sea level. You will remember that in our specifications we stated a height of between 65 and 70 metres. It will consist of 48 spans, and each span will be 850 metres in length. It will carry passenger and goods vehicles on its 4 motorway lanes, and will be 16 metres wide in total. The company estimates that it can carry 6000 cars per hour.

The second plan is for a combined structure. Here the tunnel will be at a depth of 50 metres below sea level. The bridge will be 8 kilometres long on each side with 8 spans. Each span will be 1 kilometre long. However, the total length of the structure will be the same as plan 1. The motorway will consist of 4 lanes for passenger and goods vehicles, and will be 18 metres in width. The construction company estimates that it will carry 3000 cars per hour in one direction.

Moving on to plan number 3. If you look at your diagrams, you will see that the length of the planned structure is greater than the other constructions. The reason for this is that the entry and exit points will be approximately 6 kilometres inland on each side. However, with an estimated capacity of 14000 vehicles per hour, it can carry much more traffic than its competitors.

The fourth plan is again 36 kilometres long. It will have 6 motorway lanes — 1 for slow-moving vehicles and the other 2 for overtaking — and will carry an estimated 6000 vehicles per hour in one direction. In contrast with plan number 1 the structure will consist of only 7 spans, each 5 kilometres long. As you can see from your plans it will be built at a height of 70 metres above sea level.

Our final plan, number 5, is very similar to number 2. The main differences are the width and the capacity. The advantage of this structure is that the company estimates that the motorway, which will be 26 metres wide, can carry a total of 8000 vehicles per hour.

That concludes my brief description of the 5 plans and I suggest that we . . .

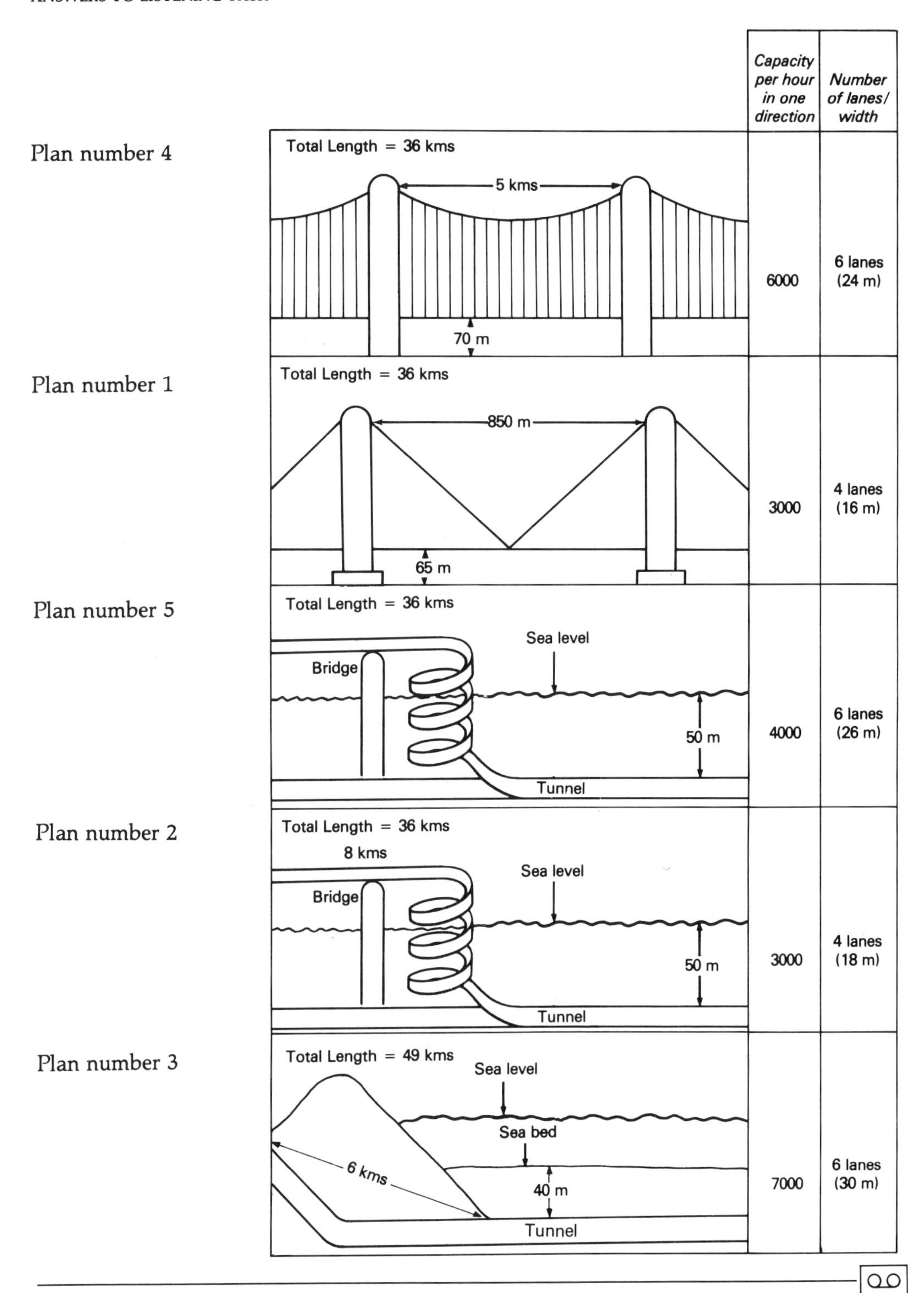

	Capacity per hour in one direction	*Number of lanes/ width*
Plan number 4	6000	6 lanes (24 m)
Plan number 1	3000	4 lanes (16 m)
Plan number 5	4000	6 lanes (26 m)
Plan number 2	3000	4 lanes (18 m)
Plan number 3	7000	6 lanes (30 m)

3 Controlled Practice

1. at a depth of
2. 10 metres
3. length
4. width
5. 4 metres high/in height
6. at a height of
7. 8 metres
8. long/in length
9. 3.5
10. width
11. height
12. can/will carry

4 Transfer

Student B: You are a civil engineer. You are going to hear your partner's plan for a single storey office block. As your partner describes the plan and its specifications, draw it on the graph below. After you have completed your diagram, compare it with your partner's.

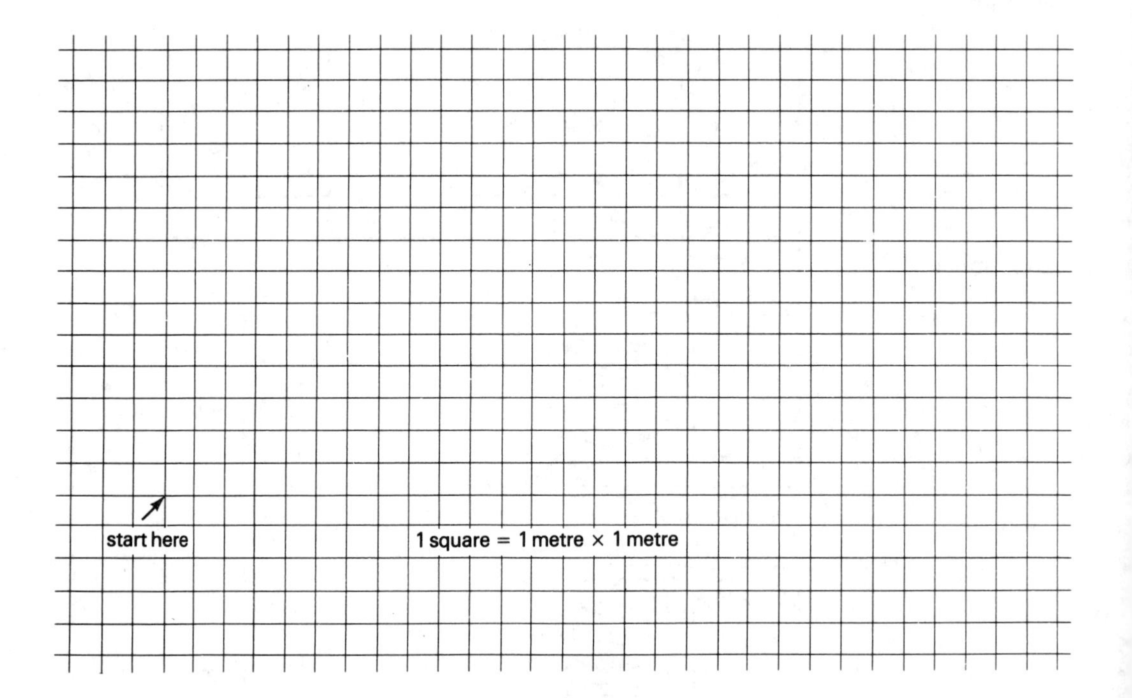

Student B: Now you have designed a plan for a 2 storey office block. Describe the plan and its specifications to your partner according to the diagram below. Your partner will draw it on his/her graph. After your partner has completed the plan, compare diagrams.

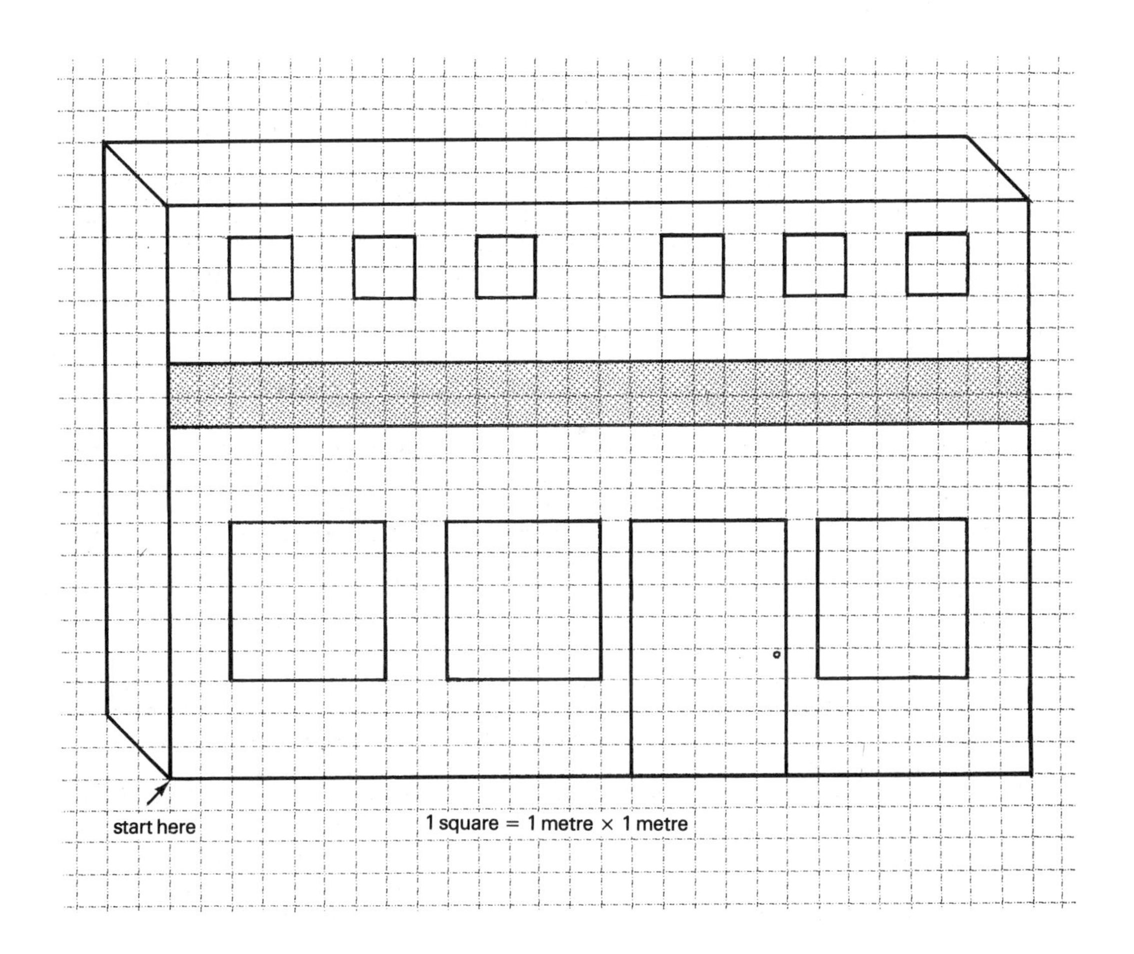

UNIT 2 Protecting your Computer System

1 Listening

TAPESCRIPT

COMPUTER CONSULTANT: This morning I would like to find out what your company requires in terms of protection for your computer system. After you've outlined what you need, then we can discuss the actual details.

DATA PROCESSING MANAGER: Well, as you know, we're a medium-sized manufacturing company. We haven't got a computer room; the main processing equipment is housed in the general office. We've got 10 terminals around the building. Four of these are used by keyboard operators to input information, and the others are used to provide information to senior staff.

COMPUTER CONSULTANT: So you haven't got a specially designed computer room?

DATA PROCESSING MANAGER: No.

COMPUTER CONSULTANT: Well, of course it's more difficult to totally protect your system in that case.

DATA PROCESSING MANAGER: Yes, I understand. However, our main need is that the system must only allow authorized access. Unfortunately in our case I don't think that it will be feasible for us to provide physical isolation. We just can't provide a separate room for the hardware. But in my opinion, we needn't provide total protection for hardware and software.

I see the solution as follows: The system must tell us if someone – an unauthorized person for example – tries to get into the system. In this way we can protect the hardware in an open environment.

COMPUTER CONSULTANT: I see, yes. I follow your logic, and I think we can design a control system to suit your specifications. But I need to know more about control of access.

DATA PROCESSING MANAGER: Yes, certainly. We can specify the *type* of access allowed to different categories of users. Firstly, the company employees don't all need to have access to the system. And secondly, some staff don't need to use all the facilities of the system. The system must only allow users to gain access via a unique identification code. And the type of identification code will enable the user to view different types of information. The system, for example, must allow managers to call up data relevant to their needs: but it may not give them access to data intended for different categories of personnel.

Now ... the next feature of the system is that it must control the *level* of access via passwords. What this means is that the system must control the type of operations which different categories of personnel may carry out on the data. Managers must be able to view, enter and amend data; operators, on the other hand, must not be able to make any changes to data – they may only make new entries. Now, in order to avoid accidental or intentional loss of data, only specially authorized personnel may delete data files.

ANSWERS TO LISTENING TASK

Security checklist *Specifications*	*Necessary*	*Unnecessary*	*Permitted/ possible*	*Not permitted/ impossible*
only allow authorized access to system	√			
provide physical isolation for hardware				√
protect both hardware and software		√		
indicate unauthorized access to system	√			
protect hardware in an open environment			√	
all company personnel have access		√		
all members of staff use all computer facilities		√		
control access via identification codes	√			
control level of access via passwords	√			
allow managers to enter, view and amend data	√			
allow operators to amend data				√
only authorized personnel delete files			√	

3 Controlled Practice

1. A keyboard operator may/can amend the Name & Address File.
2. A senior manager needn't/doesn't need to update the N & A File.
3. A keyboard operator may not/can't/mustn't delete the N & A File.
4. Any authorized user may/can call up the N & A File.
5. A keyboard operator may not/can't amend the Budget Forecast.
6. A company executive may not/can't delete the Budget Forecast.
7. Computer operators needn't/don't need to switch off the system in the evening.
8. A computer operator must store away all file disks in the evening.
9. A computer operator must power up the system in the morning.
10. A keyboard operator needn't/doesn't need to check the Supplier File.
11. A/The Purchasing Manager must check the Supplier File before placing an order.
12. A keyboard opertor may not/can't amend the Debtor File without the approval of the Account Manager.

UNIT 3 Designing a Computer System

1 Listening [cassette icon]

TAPESCRIPT

PAUL BAILEY: Yes, well I'm not a computer expert. Could you please explain what this chart means?

COMPUTER CONSULTANT: Yes, of course. A computer system consists of hardware and software. Hardware means the different types of equipment or devices, and software means the programs. OK so far?

PAUL BAILEY: Yes.

COMPUTER CONSULTANT: Anyway, hardware comprises 2 components — the central processing unit and the peripheral equipment. The central processing unit or CPU is the brain of the computer. It controls all the equipment and processes information — just like your brain controls your body and also makes decisions. The other component, peripheral equipment, we use for 3 purposes — firstly to give information to the computer as input; secondly to store information for the future; and thirdly to get information from the computer as output. So we can split peripheral equipment into input, storage and output. Now let's take the payroll as an example. What do you want to get out of the computer?

PAUL BAILEY: The wages of my employees.

COMPUTER CONSULTANT: Uh-huh. So that's your output. Now in what form do you want this information to appear?

PAUL BAILEY: On paper.

COMPUTER CONSULTANT: Right. So on your diagram under output we've got a printer. A printer is one kind of output device and of course it prints information. Also on your diagram is another type of output device — a visual display unit or VDU. Now a VDU is a kind of screen. Often you don't want to print everything — you just want to see what's what. So we can classify a printer and a VDU as peripheral output equipment. OK so far?

PAUL BAILEY: Yes.

COMPUTER CONSULTANT: OK. Let's go on a bit further, but let's stay with the payroll. What kind of information do you need in order to calculate the payroll?

PAUL BAILEY: Employee's name, hours worked, hourly rate, overtime worked, overtime rate, and tax code. And then I need a calculator and tax tables.

COMPUTER CONSULTANT: Well, this brings us back to the other 2 categories of peripheral equipment — input and storage. But let's consider storage first. A store is like a file. At present your files consist of sheets of paper or card. With a computer system your storage devices fall into 2 categories. You'll see them on your diagram under storage — tape and disk. They can contain information about employees just like a file, and they operate like a tape recorder or record player. You can put information on and record it, and you can get that information back at a later time — and of course you can change the information very easily. And finally let's consider your input device. The simplest and most common type of input device is a keyboard. It consists of a typewriter keyboard connected to the other parts of the system. For your payroll calculation you load your employee file from a tape or disk machine. Then the computer will find your first employee and show you his details on the VDU. Then it will ask you to input the details of hours worked, etc. And finally it will calculate his or her wage and print it onto the printer. The only thing we haven't mentioned is software. Software means programs, and a program consists of a set of instructions which tell the computer how to process the information. For a payroll calculation the program tells the computer what to multiply, what to add, what to subtract, etc. — in fact how to do the whole calculation. And when you are not using a program, it is stored on tape or disk — just like your files.

PAUL BAILEY: OK. Enough theory for the moment. When can I see one of these wonderful systems?

COMPUTER CONSULTANT: I'm glad you asked me that because by chance I've got a beautiful little computer in the back of my car. I'll just go and gt it.

ANSWERS TO LISTENING TASK

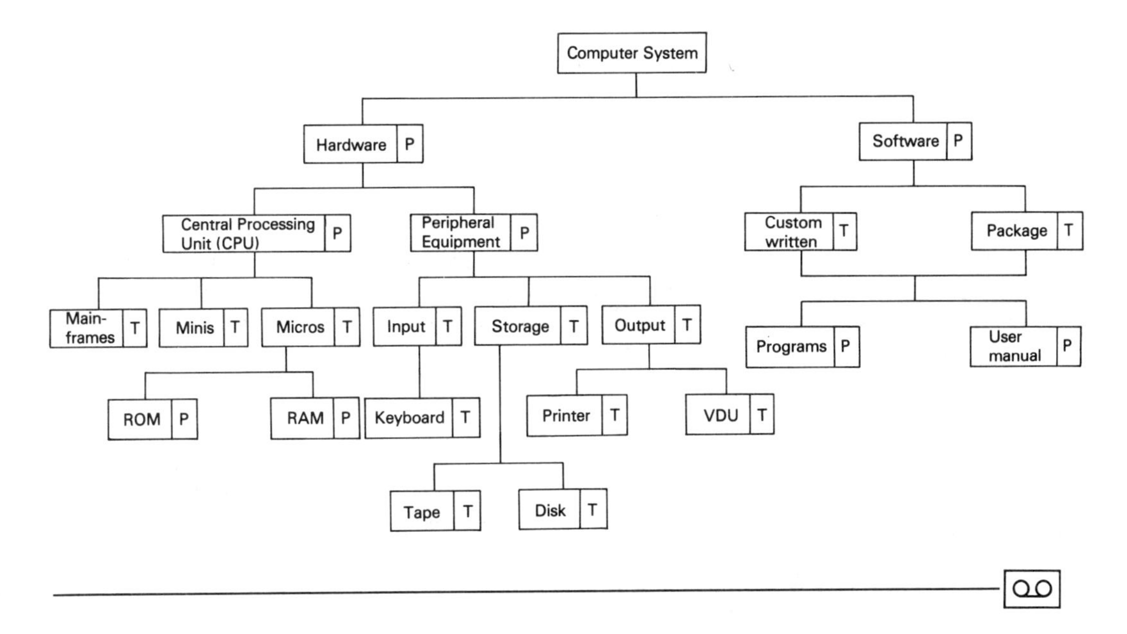

3 Controlled Practice

A. See the chart above.

B. 1. A computer's hardware *comprises* a CPU and peripheral equipment.
2. Input, output and storage devices are *types/kinds of* peripheral equipment. Storage devices *fall into* 2 categories — tape or disk.
3. We can *classify* a printer and a VDU *as* output devices.
4. We *can divide* software *into* 2 types — custom written and package.

4 Transfer

Student B: You are going to hear the description of the ABC Model A computer. As your partner describes the computer, complete the chart below. After you have completed your chart, compare it with your partner's.

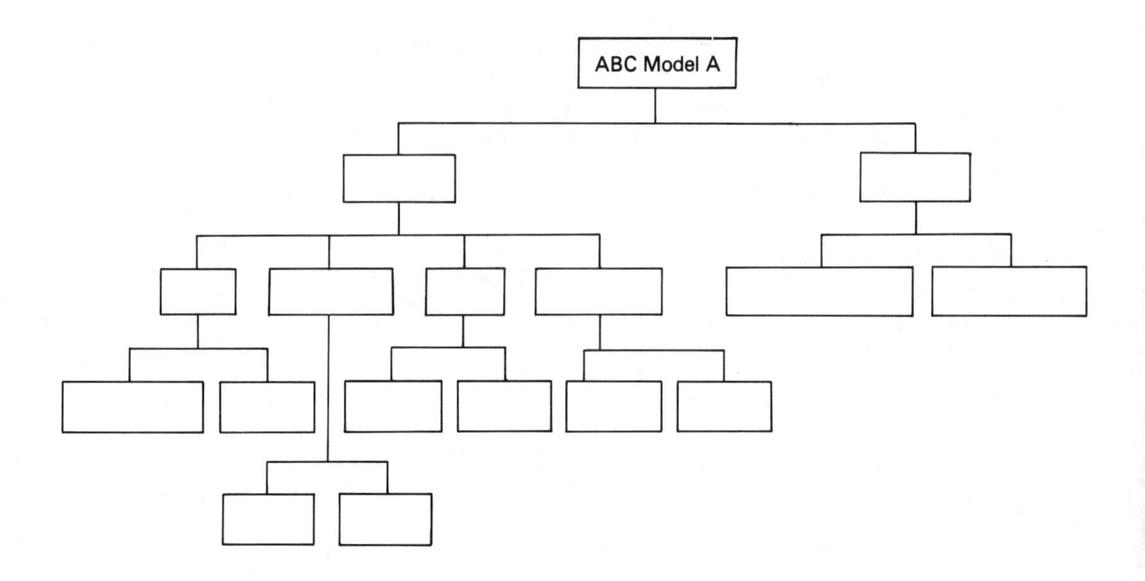

Student B: Now you are going to describe the ABC Model B computer to your partner. You will find the details in the chart below. Your partner will draw a chart according to the details you give. After your partner has completed the chart, compare your versions.

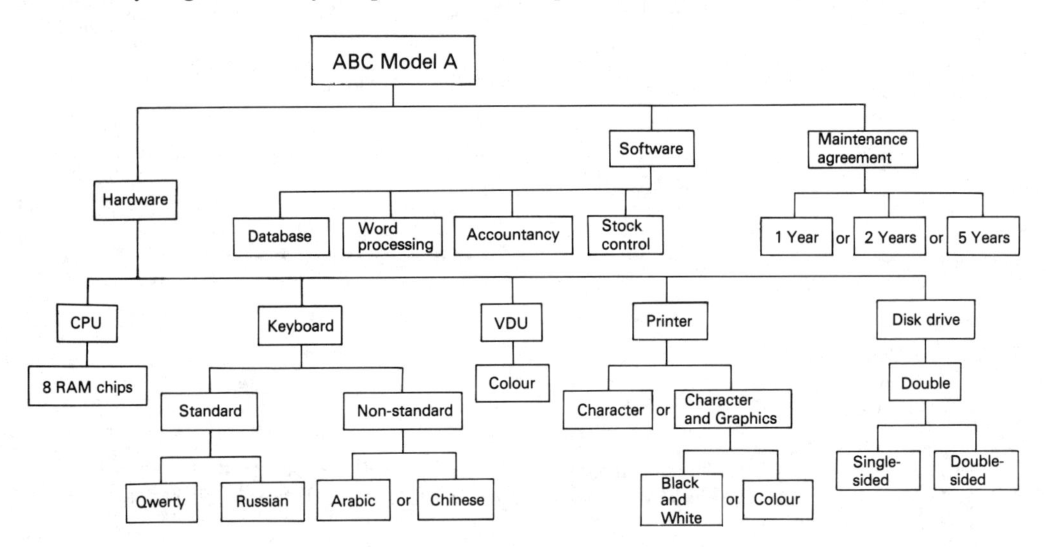

UNIT 4 Optical Fibres

1 Listening

TAPESCRIPT

ENGINEER: There's no doubt optical fibre systems have enormous advantages over existing transmission systems.

JOURNALIST: So we have heard. But what *are* these advantages?

ENGINEER: Well, first and foremost, they have a much higher capacity than copper wires. In other words, they can carry much more information — telephone calls or data, it doesn't matter which.

Secondly, they have a potentially lower material cost. At the moment, production costs of optical fibre are relatively high, but this is only because mass production hasn't really started. In the long term, optical fibre will cost much less to produce than conventional cables. Um, . . . another big advantage is their size . . . they take up much less space. With conventional cable you need many more ducts or pipes to carry the cable.

JOURNALIST: What about quality of transmission?

ENGINEER: Yes, they score very highly here as well. The signal doesn't need to be amplified as often as with conventional cable, where signal loss means you need far more repeaters or boosters — in fact, every 2 or 3 kilometres as opposed to every 20 kilometres.

Also, as far as quality is concerned, optical fibres don't suffer from interference or crosstalk as much as conventional cable.

JOURNALIST: Anything else?

ENGINEER: Yes, they also have complete electrical isolation and therefore there's much more security on the line — the data can't be corrupted or interfered with.

ANSWERS TO LISTENING TASK

1. d 2. c 3. e 4. a & f 5. b

3 Controlled Practice

A.

1. C
2. C
3. U
4. U
5. C
6. U
7. U
8. U
9. U
10. U
11. U
12. U

B.

1. much/far more
2. many/far more
3. much/far less
4. much/far less
5. much/far more
6. much/far less
7. much/far fewer
8. much/far less
9. much/far more
10. much/far more

4 Transfer

Some of the advantages could be:

size, cost, efficiency, reliability, flexibility, capacity, etc.

UNIT 5 Shape Memory Alloys

1 Listening

TAPESCRIPT

Shape memory alloys have many applications, but before I give some examples, what exactly is a shape memory alloy?

Essentially it's a metal which can be deformed when cold and will remember its original shape when heated. The particular alloy we are using here in the lab is nickel titanium.

We can see one application here in a conventional piston. When the piston is cold, the SMA coil or spring contracts and so the piston doesn't move. Heat causes it to expand and consequently the piston moves up. The advantage is that the device can work without any mechanical power, just from the heat which is supplied by the engine itself.

Let's look at some other applications. Over here, we have a domestic coffee machine. We've fitted a valve controlled by an SMA actuator. When the coffee machine reaches a certain temperature, the SMA actuator opens the valve and so the water is dumped onto the coffee. Again the system works without any additional power – electrical or mechanical – just the heat of the steam given off as the water heats up.

One more application – this time for the motor industry. We're experimenting with a spring device which *expands* when it cools. This device will be fitted to the cooling fan. In cold weather, the spring expands and this makes the fan close down, and thus helps to warm up the car. In warm weather, the spring contracts and therefore causes the fan to operate and, as a result, helps to cool down the car. Again a system using no external or additional power source. Right, are there any questions?

ANSWERS TO LISTENING TASK

Application	*Cause*	*Primary effect*	*Secondary effect*	*Tertiary effect*
1. Piston	Piston cold	Coil contracts	Piston doesn't move	
	Piston hot	Coil expands	Piston moves up	
2. Coffee Machine	Reaches certain temperature	SMA actuator opens valve	Water is dumped onto coffee	
3. Cooling fan in car	Cold weather	Spring expands	The fan closes down	Car warms up
	Warm weather	Spring contracts	The fan operates	Car cools down

3 Controlled Practice (other versions are possible)

A. 1. When the coffee machine reaches a certain temperature, the SMA actuator opens the valve and consequently the water is dumped onto the coffee.
2. When it is cold, the spring expands. This makes the fan close down and so the car warms up.
3. In warm weather, the spring contracts and therefore the fan operates. This causes the car to cool down.

B. When the temperature rises, the metal expands and consequently the contact is made. This causes the heater to be switched off and so the temperature drops. In time, the metal contracts and thus the contact is broken. The broken contact causes the heater to be switched on and so the cycle starts again.

UNIT 6 What is a Transistor?

1 Listening

TAPESCRIPT

Transistors nowadays are made of silicon. So before we look at an actual transistor, what is silicon?

Well, after oxygen, silicon is the most common element on this planet. This is obviously important since it doesn't cost much to get the raw material. It's a solid and, as such, has the ability to conduct electricity. Let's look at this other table (Figure 2) which shows resistance of a number of different elements. One from the top is polythene — it's very resistant to electricity: it doesn't allow it to flow and therefore is used for insulating. You see this piece of copper wire is insulated with polythene. At the bottom of the table are materials of very low resistance such as iron and copper — these are very good conductors — they allow electricity to flow and are therefore used for making electric cable. Now, silicon is right in the middle, in other words neither a good conductor nor a good insulator — that's why it, and other solids such as germanium, are known as semiconductors.

As a semiconductor, silicon is very sensitive to impurities — these are called dopants. If you add as little as 0.0001% of a dopont to silicon, you can increase conductivity by 1000 times. So, now let's look at this schematic diagram of a transistor. As you can see it consists of two types of silicon. One type has been doped with boron which gives it a positive charge — this is termed p-type silicon. The other type has been doped with phosphorus which gives it a negative charge — this is known as n-type silicon. By doping a silicon crystal with p and n type dopants a p-n junction is formed. So, here in this diagram you've got a substrate of p-silicon and then two pockets of n-silicon. There are three contact leads — the one on the right is connected to the metal source. When a voltage is applied to this contact, a current flows through the p-n junction and out through the metal drain — this contact on the left. In the middle you have another electrode — the gate which can be used to regulate the current flow. If we look at . . . *(Fade)*

ANSWERS TO LISTENING TASK

Fig 1 a. Oxygen
b. Silicon

Fig 2 c. Polythene
d. Silicon
e. Semiconductors
f. Copper
g. Conductors

Fig 3 h. n-type silicon
i. n-type silicon
j. metal drain contact
k. gate contact

3 Controlled Practice

A. 1. consist of/are composed of/
   are made of/are formed from
2. is used for
3. consists of/is composed of/
   is made of/is formed from
4. are . . . called/known as/termed
5. allow . . . don't allow
6. At the top
7. In the middle
8. On the left
9. On the right
10. At the bottom

B. 1. called/known as/termed
2. is used for
3. consists of/is composed of
4. has the ability
5. can
6. allows

4 Transfer

Student B: Ask your partner questions about two objects. Guess what they are. Ask the following questions:

What is it?
What can it do?
What is it used for?
What is it composed of?

NOTE: Do *not* ask what it is called. You must decide from the description.

Then describe the following to your partner in answer to his/her questions: a fuse, an escalator.

UNIT 7 Testing Circuits

1 Listening

TAPESCRIPT

I believe you know a little about our Logic Analyser, the IMAT, but before I demonstrate it, let me give you one or two important facts. First of all, what are its applications? It is designed as a general-purpose tool for circuit analysis — that is circuits which are either being developed, in production or need servicing. There are 6 basic modes of operation, but today I'll be concentrating on just one — the most common one — that is the 16 channel configuration used for normal measurements in digital circuits. In this mode we can sample at a frequency of 50 MHz (megaherz).

OK, let's look at the front panel. On the left hand side, we've got the screen and, very important, 8 softkeys located immediately below the screen. These softkeys are used to set up the IMAT. So, if we press this button here — just to the right of the screen, at the top — for Mode Selection, we can choose 16 channel timing. You can see the basic menu on screen and the functions of the softkeys in this mode. So, now we can set it up quickly. Let's first enter 'clock' with this softkey on the left and you'll see the menu for 'clock' as it's now set up. Press 'menu call' — the button next to 'mode selection' — and we're back to the basic menu. Let's just enter some data on the 'threshold' — press the threshold softkey and then enter the threshold you want — say 1.6 volts — using the 'data entry' keys — here in the centre to the right of the screen.

OK, of course you can change the other parameters, but let's leave it as it is set up, and connect up this probe to a microprocessor circuit — here you can see it's very easy on the back panel.

If we want to see a 'Timing display', we must carry out the following procedures: we start by pressing the button here in the top, right-hand corner — 'Start'. Then we press this button here, just below 'Memory', marked 'mem' — I'll tell you more about that later — so, we press this button here marked 'Time Display'. Now, you see on screen, 8 of the channels, displayed from A0 to A7. They are magnified 10 times but we can increase the magnification, if you want, by pressing one of the softkeys. Down here the two softkeys at the end on the right, labelled 'up/down' can be used to magnify 'down' or 'up'. If we want to look at some other channels, we just scroll up or down using the first two softkeys on the left — you see it's marked here 'Scroll' down or up. At the moment we're looking at a display range from 252 to 352 N/sec (nanoseconds) but you can change the position of the window in the main memory by moving the shift select here in the centre of the softkeys with this softkey on the left marked 'Expand'. Just one thing to remember each time you press a softkey — press this return button just to the right of the screen at the bottom of the set. Well, would you now like to try it yourself?

ANSWERS TO LISTENING TASK

1. **Basic Menu**
   a. softkeys (8)
   b. mode selection
   c. clock
   d. menu call
   e. data entry

2. **Time Display**
   a. start
   b. time display
   c. up
   d. scroll
   e. expand
   f. return

3 Controlled Practice

OK, *let's* look at the back panel. You *'ll/can see* that there are two probe connectors *at the* top *on the left-hand* side. *Below* the two probe connectors, there's an IEC-bus connector. *If* we connect this, we can transfer all the data to another system. Right, *let's look* at the input and output sockets *at the* bottom of the *back* panel. The first one *on the* left is for input of a counter and signature analyser. Next to this is the analog recorder input. *In the* centre is the video output – we can display data on another screen *by* connecting a video terminal. *To the* right of the video output are two other sockets: one for a link line, the other for an external qualifier.

Of course, *just one thing to* remember, *don't* forget to connect the power supply to the three pin socket *in the top right-hand* corner!

UNIT 8 Setting up your new Computer

1 Listening

TAPESCRIPT

Welcome to your new QD Personal Computer. Look in the box and you will find all the parts to set up your system. But you will need to use your own TV. First take out your QD computer and look at the back, where you will find 4 sockets. The package also contains an aerial lead. You will need that now. Connect one end of the aerial lead to the TV socket and the other to your TV. If your TV has 2 aerial sockets – UHF and VHF – don't use the VHF one. Now the power supply unit, which is also provided in the package. At the back of the computer you will see a socket marked DC-in. Connect one end of the supply unit to this DC-in socket. Connect the other end to a mains electricity socket. Your computer is now on.

Now the micro-disk drive and the lead, which you will find in the package. The lead consists of 2 cables. Each cable has a plastic label – one with MIC (mike) and the other with EAR (ear). Connect the MIC cable to the MIC socket at the back of the computer, and the EAR cable to the EAR socket. Don't worry about which end you use – they are the same. The other ends of the leads go into the micro-drive – one to EAR on the micro-drive, and the other to MIC. Now plug the power unit into the electricity socket at the mains. OK, you're nearly ready.

Now tune in your TV. Firstly turn it on, but with the volume down. Now press a button to select a channel that you don't use for a TV channel. Now use the tuning mechanism to find the right position. When you see the message QD COMPUTER on your TV screen, it is correctly tuned in. Always use that channel with your computer, because it is always ready for action. Now your system is connected and ready for use.

Finally take the 'Welcome' disk out of the package and insert it into the front of the disk-drive. Don't use force; and make sure the disk is the right way up. Just look at the arrows. Insert the disk and press the keys SHIFT and BREAK on the computer. The rest of the information is on the disk and it will appear on the screen.

If you have any problems with the equipment in the package, don't try to repair it yourself. Take it back to your supplier.

We wish you a lot of enjoyment with your new QD Personal Computer.

ANSWERS TO LISTENING TASK

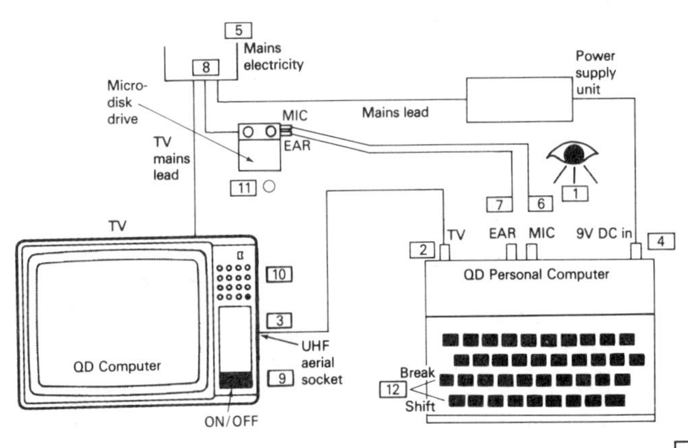

3 Controlled Practice

1. look
2. find/take
3. insert/plug/push
4. insert/plug/push
5. do
6. take/find
7. insert/plug/push
8. insert/plug/push
9. follow
10. look
11. find/take
12. take/find
13. connect
14. connect
15. don't connect
16. do
17. plug
18. plug

4 Transfer

Student B: Your partner is going to give you instructions on how to insert fanfolded paper into a printer. Number the pictures below in the correct order.

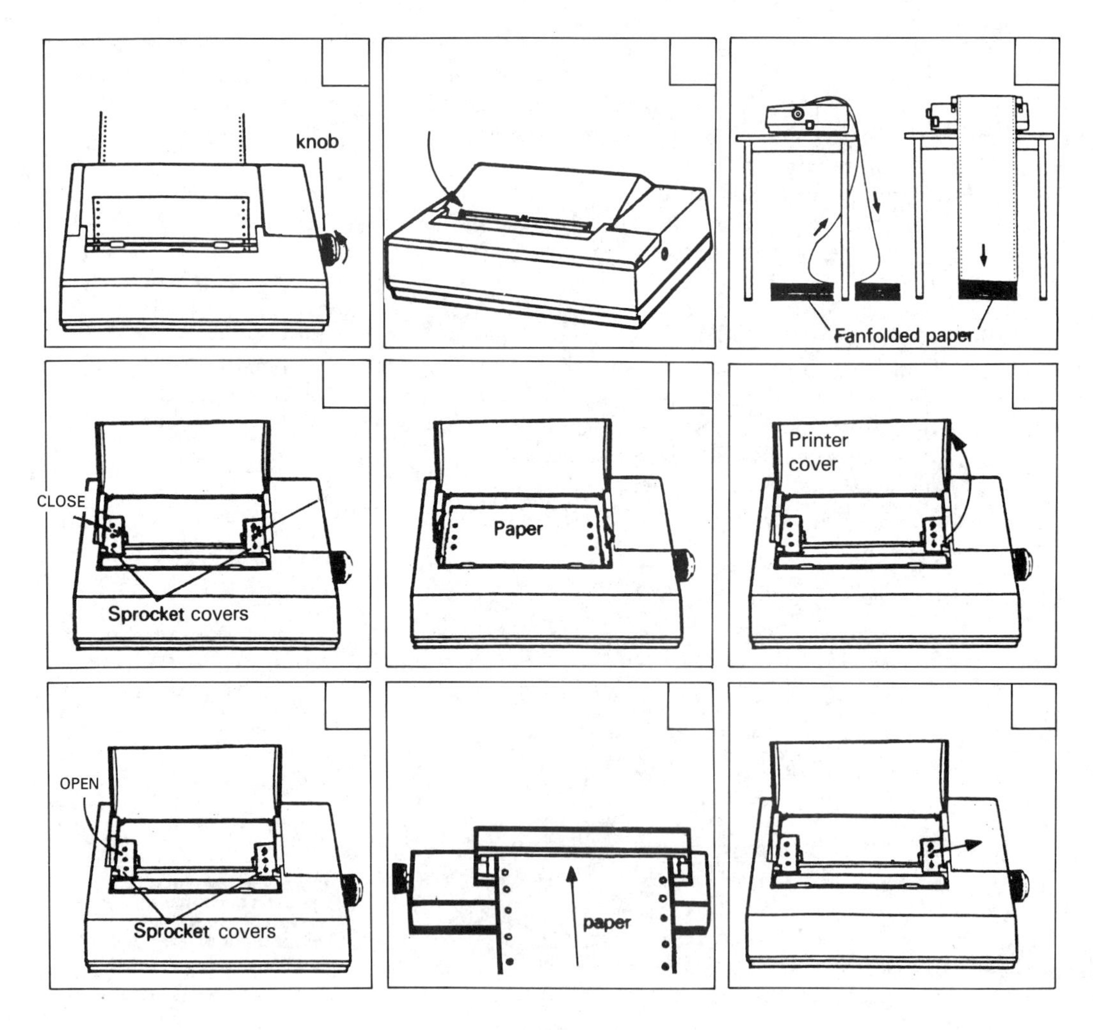

UNIT 9 Building a House – the Requirements

1 Listening

TAPESCRIPT

A house in this country needs protection against the elements, the environment and a number of other risks. Of course the location of the house is an important factor. However, some types of protection are important for both city and country houses. Other types of protection will depend on whether the house is in a city or in the country. And, of course, although protection is important, it is certainly not required against everything. In fact it would be impossible to give a house total protection. So we must look at each factor and decide how important protection is in each case.

The first factor is solar radiation. As we live in a temperate climate, it is unlikely that a house will need protection against it. However, in hot climates this type of protection is often provided by means of external wooden shutters or internal blinds. These are closed to prevent the sun from entering the house. But I think that in Britain both town and country houses are unlikely to get too much sun!

Moving on, the next factor is rain. All houses in this country are certain to need protection against rain. This type of protection is provided by solid brick walls and a sloping roof. Of course we don't have monsoons, but our climate can be very rainy and damp.

The next factor is high winds. Here I think we need to distinguish between city and country houses. A country house in an exposed location may need protection. But I think that in a city a house is unlikely to be so exposed, and therefore unlikely to need protection against high winds.

The fourth factor is noise from outside. But a normal brick structure, perhaps with double-glazed windows, provides sufficient protection. Again we need to look at the location of the house. Even in a city, there are likely to be differences. Some houses are in quiet areas; others in noisy areas. So a city house may need protection against outside noise.

Now I'd like to move on to the next factor – noise from inside. Here we need to consider how to insulate one room against noise coming from another room. For this purpose interior brick walls are recommended. Here both city and country houses are likely to need protection.

I've already mentioned rain, but there is also damp from underground or rising damp. And it is likely that both city and country houses will need protection against this. Good foundations and a solid stone floor are ideal for this type of protection.

The next factor is the risk of fire from outside. All houses will certainly need protection against this. There is also the danger of fire from inside. In this case we need to consider the type of heating used, as well as other factors. In my opinion, all houses may require this type of protection.

The next point is heat loss. Although we live in a temperate climate, the British winters are typically cold and very damp. So houses are bound to need protection against heat loss. Here again thick walls and well-fitted doors and windows prevent heat loss.

And finally there is the question of heavy snow. Heavy snow is not normal in most parts of Britain, and therefore a house here definitely won't need this type of protection.

ANSWERS TO LISTENING TASK

FUNCTIONAL REQUIREMENTS OF A CITY HOUSE

	Certainly required	*Probably required*	*Possibly required*	*Probably not required*	*Certainly not required*
Protection against solar radiation				✓	
Protection against rain	✓				
Protection against high winds				✓	
Protection against noise from outside			✓		
Protection against noise from inside		✓			
Protection against damp from underground		✓			
Protection against fire from outside	✓				
Protection against fire from inside			✓		
Protection against heat loss	✓				
Protection against heavy snow					✓

3 Controlled Practice

1. It is likely that we will hear traffic noise from outside.
   We are likely to hear traffic noise from outside.
2. I am sure/certain/positive that the house will lose/loses heat.
   The house is certain/bound to lose heat.
   The house will certainly/definitely lose heat.
3. The house definitely/certainly won't become too hot.
   I am sure/certain/positive that the house won't become too hot.
4. Damp may/might rise into the house.
5. It is unlikely that rain will come through the roof.
   Rain is unlikely to come through the roof.
6. We definitely/certainly won't feel wind through the walls.
   I am sure/certain/positive that we won't feel wind through the walls.
7. It is likely that fire will spread inside the house.
   Fire is likely to spread inside the house.
8. The roof definitely/certainly won't collapse because of snow.
   I am sure/certain/positive that the roof won't collapse because of snow.
9. I am sure/certain/positive that we will hear noise from other parts of the house.
   We are certain/bound to hear noise from other parts of the house.
   We will certainly/definitely hear noise from other parts of the house.
10. Fire may/might spread from outside the house.

UNIT 10 Installing the 9450 Photocopier

1 Listening

TAPESCRIPT

Now the new 9450 model is very simple to install. The box contains all the parts except the cabinet. This is an optional extra. Now, before you visit a customer you must check if he wants the cabinet or not. You shouldn't presume he doesn't just because it isn't on the order form. As you know, many customers don't realise exactly what they want or need. And, of course, you ought to try to persuade him to take the cabinet.

Now, on to the installation. The box contains 5 parts – the main copier, the lid, an A4 paper cassette holder, an A3 paper cassette holder, and a copy tray to receive the finished copies. It also contains 6 bottles of toner for the 9450. You should assemble the copier in its final position – either on a desk top or on the cabinet. You shouldn't use a surface which is either too small or too low. And I emphasise that you are not to assemble the machine anywhere except in its final operating position. First, you are to take out the main copier and adjust the feet, so that the machine is level. Next, you should attach the lid. You are not supposed to use force – it just clips on. Next you ought to slide in the A4 cassette holder into the upper space. Again I emphasise that it is not to go into the lower space. You are supposed to just click it into position. Then you ought to fit the A3 cassette holder into the lower space. Finally the copy tray is to be fitted on the opposite side. You oughtn't to need any force – it just sits in position. Now the copier is assembled. Next take 2 bottles of toner. You mustn't use any other toner except 9450. The bottles must be shaken first, and then poured into the toner compartment.

Switch on the machine, and it should be ready for use.

ANSWERS TO LISTENING TASK

Instruction	*Do's*	*Don'ts*
check if the customer wants a cabinet	√	
presume he doesn't want a cabinet because it isn't on the order form		√
try to persuade the customer to buy a cabinet	√	
assemble the copier on the floor		√
assemble the copier in its final position	√	
adjust the feet so that the machine is level	√	
use force to attach the lid		√
slide the A4 cassette holder into the lower space		√
fit the tray under the A4 cassette holder		√
pour 2 bottles of 5100 toner into the toner compartment		√

3 Controlled Practice

A.

You must push the reset button.	9
You shouldn't use force when replacing the drum.	8
You oughtn't to try to pull out the copy drum before you release the locking screw.	3
You must switch the power off.	1
You ought to replace the copy drum carefully.	7
You ought to check the machine is functioning.	12
You should close the front panel.	10
You mustn't forget to switch the power on again.	11
You should remove the jammed sheet of paper.	5
You should open the front panel.	2
You are not (supposed) to touch the drum surface.	4
You mustn't use a sharp object to remove the sheet.	6

B.
1. You *must/should* wait until the machine is ready to copy.
2. You *ought to* press the copy key now.
3. You *are supposed* to add paper.
4. You *mustn't/shouldn't* use the machine before you add some toner.
5. You *oughtn't to* try to press the copy key as the machine is jammed.

4 Transfer

PAIRWORK

Student B: You are the Technical Manager of a video shop. You receive a telephone call from Student A, who has just bought a video cassette recorder from your shop. Unfortunately the installation and operation instructions for the machine are missing. Use the pictures and the notes below to tell Student A how to set up the machine. Expand the notes to give instructions.

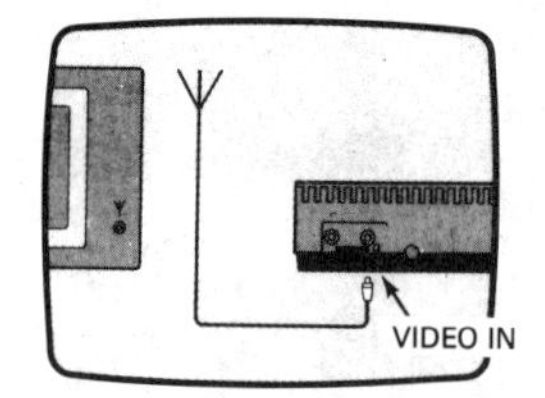

Remove aerial cable from TV/plug into 'VIDEO IN' at back of VCR.

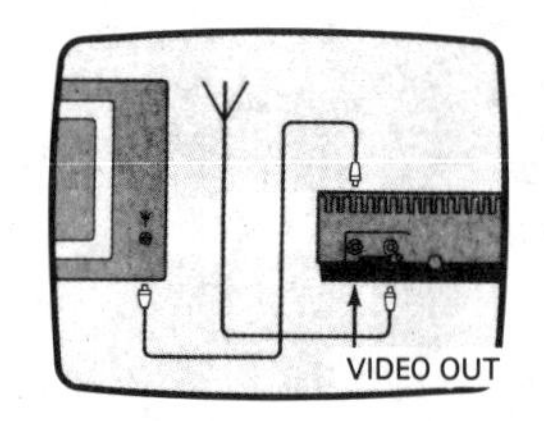

Take VCR aerial cable from box/plug one end into 'AERIAL' in TV and other into 'VIDEO OUT' on VCR.

(3)

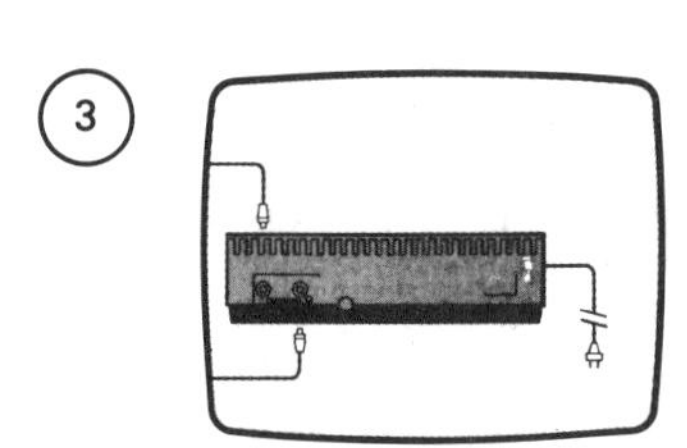

Plug VCR into mains.

(4)

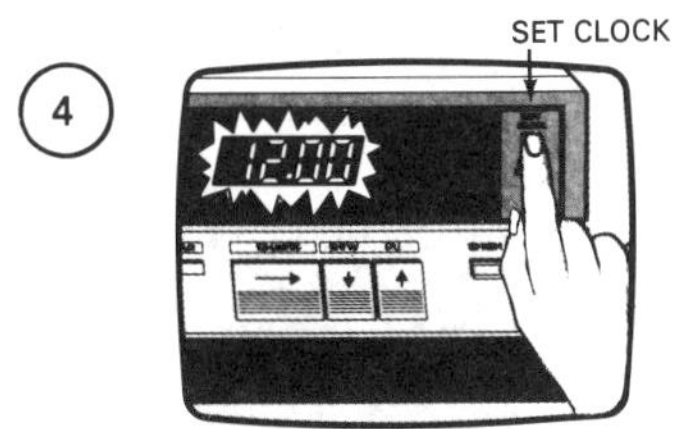

Press 'SET CLOCK' on front of VCR/clock should flash '12.00'.

(5)

Press 'UP' or 'DOWN' button to set correct time.

(6)

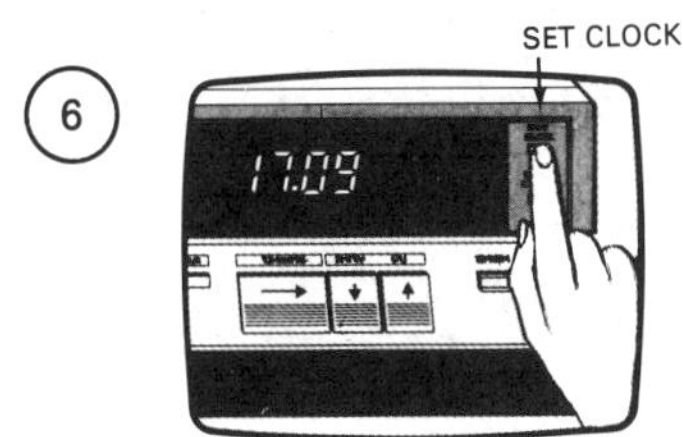

Press 'SET CLOCK' again to stop flashing and set correct time.

(7)

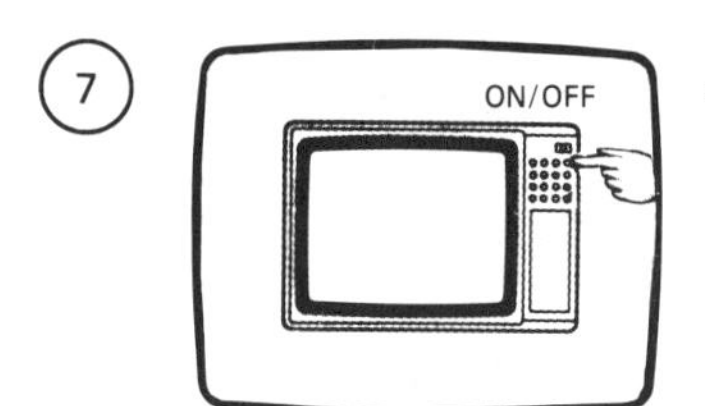

Turn on TV.

(8)

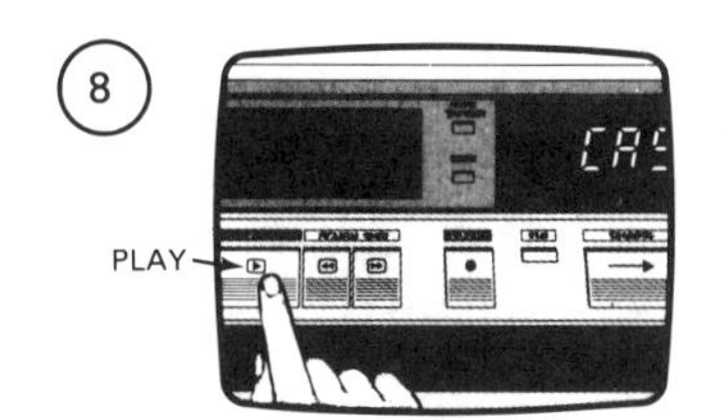

Press 'PLAY' button on VCR/do not put in a video cassette.

(9)

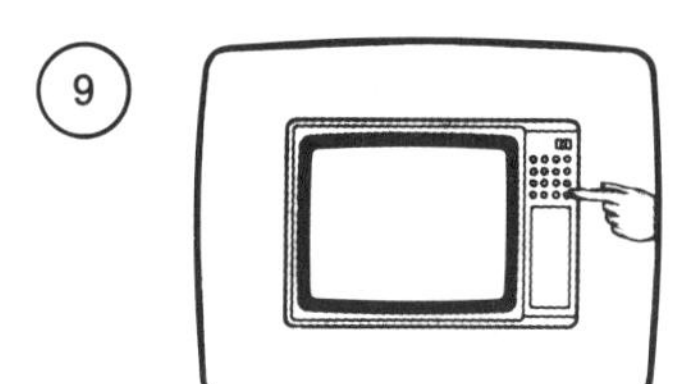

Press button on TV for VCR.

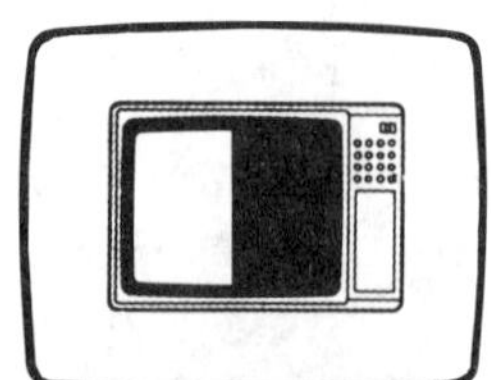

Tune in channel until you see a black and white pattern on screen.

11

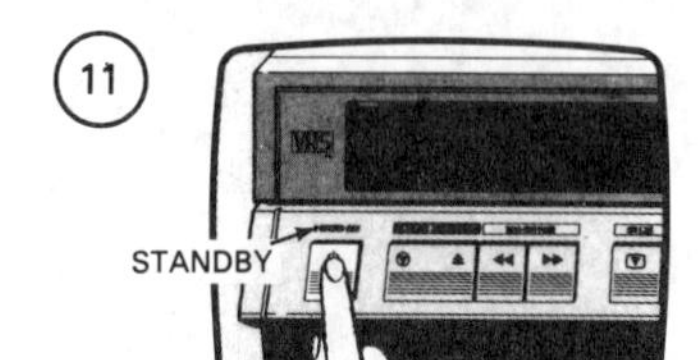

Press 'STANDBY' on VCR.

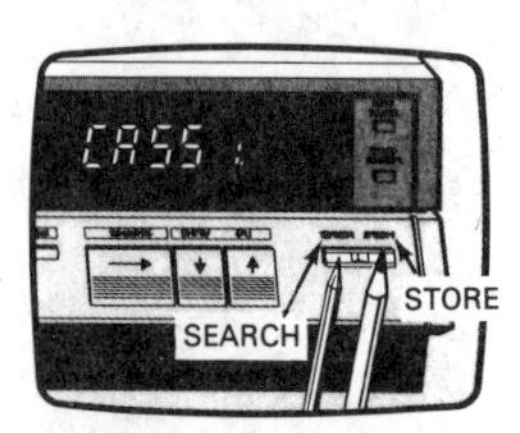

Press 'SEARCH' button on VCR to find a TV station. First look for TV1/compare TV1 picture with normal station setting on TV/assign number '1' to this setting by pressing 'STORE' on VCR.

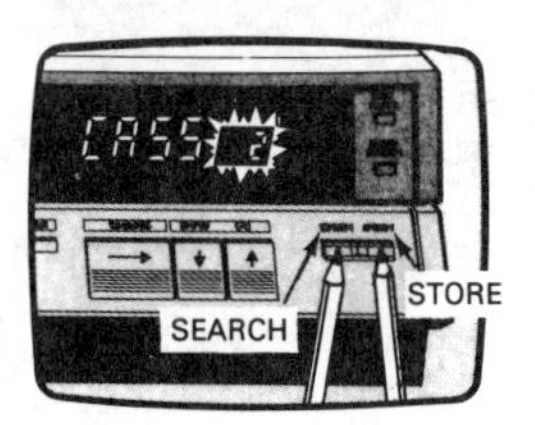

Press 'SEARCH' button to find TV2/follow same steps as above/repeat until you have stored all TV stations.

After you have given the instructions, Student A will repeat what he/she must do to check that he/she has understood.

UNIT 11 Printing Processes

1 Listening

TAPESCRIPT

In the past, the journalist typed his story on any old manual typewriter and then the story was cast in metal by a compositor. This was followed by another slow and laborious step – the type, that is the metal type, was checked and corrected before the pages were made up in metal by the printers. Finally the completed page was cast in metal ready for the printing press.

Well now we're at a sort of interim stage – not completely computerised but getting there – the journalist still types his story on a typewriter – it might be an electronic one! But the story is then re-typed by a compositor into the main computer. The computer sets the type photographically onto paper before the individual articles are pasted onto the pages. Finally the pages are photographed and plastic plates are produced for the press.

The next step, as you've probably guessed, is to computerise the whole process. The journalist will type his story onto a computer terminal – sometimes called a VDU (Visual Display Unit) and this will be sent directly to the computer.

Once again the computer will set the type onto photographic paper before the page is designed and made up on the terminal by journalists. Finally the computer will produce the page and plastic plates will be made ready for printing.

Well, that's what's planned. Any questions?

ANSWERS TO LISTENING TASK

3 Controlled Practice

A. 1. The journalist typed the story on a manual typewriter.
2. The story was set in metal by a compositor.
3. The metal type was checked and corrected.
4. The pages were made up in metal by a printer.
5. The complete pages were cast in metal ready for the press.

B. 1. The journalist types the story on a typewriter.
2. The compositor retypes the story into the main computer.
3. The computer sets the type photographically onto paper.
4. Individual articles are pasted onto the page.
5. The pages are photographed and plastic plates are produced for the press.

C. 1. The journalist will type the story on a VDU/Terminal.
2. The story will be sent directly to the computer.
3. The computer will set the type onto photographic paper.
4. The pages will be designed and made up on terminals by the journalists.
5. The computer will produce the pages and plastic plates will be made ready for printing.

UNIT 12 Energy

1 Listening

TAPESCRIPT

The picture of electricity generation in our region is not particularly rosy. If we look at the statistics over the last 10 years we can see an increase during the first 6 years of the period; and then a steady decrease over the last 4 years. However, let's look at the figures in more detail.

In year 1, when the new region was established, capacity stood at 10900 megawatts. This was to provide electricity for domestic and industrial users. This figure increased in the 2nd year by 300 megawatts, giving us a capacity of 11200 megawatts. This represented a moderate rise in line with national trends. However, in the 3rd year we did not manage to increase capacity, and it remained constant at 11200 megawatts. This was as a result of the increase in electricity tariffs for consumers. In response to this price increase, local industry introduced measures to conserve energy. The 4th year saw an improvement, with a rise of 800 megawatts for the region. This was, in fact, quite a substantial increase, and was mainly caused by a number of new domestic users. We looked forward to a continued rising trend. The trend did continue, but in year 5 capacity only went up by 200 megawatts to 12200 megawatts. Again the price rises led to a policy of energy conservation. And with the high rate of inflation that year, many users took steps to reduce their electricity consumption. The following year, year 6, we reached our peak, and capacity rose to 12400 megawatts. This was our high point. It was a very good year for the region's economy generally. However, since year 6 we have registered a steady decrease in our capacity for the region.

In year 7 we saw a drop to 12000 megawatts – which, in fact, represented a fall to the level of year 4. You will, no doubt, remember that that was the year in which our local industry began to suffer from the recession. Now, this downward trend continued in year 8 – but at a much more dramatic rate. Many factories closed down and the general economic climate was most unhealthy. And our capacity went down by 1000 megawatts in that year. This was our worst year . . . our biggest single drop. In the following year – year 9 – we managed to keep capacity at almost the same level, but we saw a small decrease of 200 megawatts in capacity. Despite our local problems, many people were beginning to realise that there was no short-term solution to the situation. And they began to spend more money generally . . . but, unfortunately, not on energy. Now, that almost brings us up to date with the 10 year review. Our final figure for the period indicates that the general decrease is still continuing; and with a further drop in capacity to 10300 megawatts for this year, I expect this downward trend to continue for at least another 2 years.

ANSWER TO LISTENING TASK

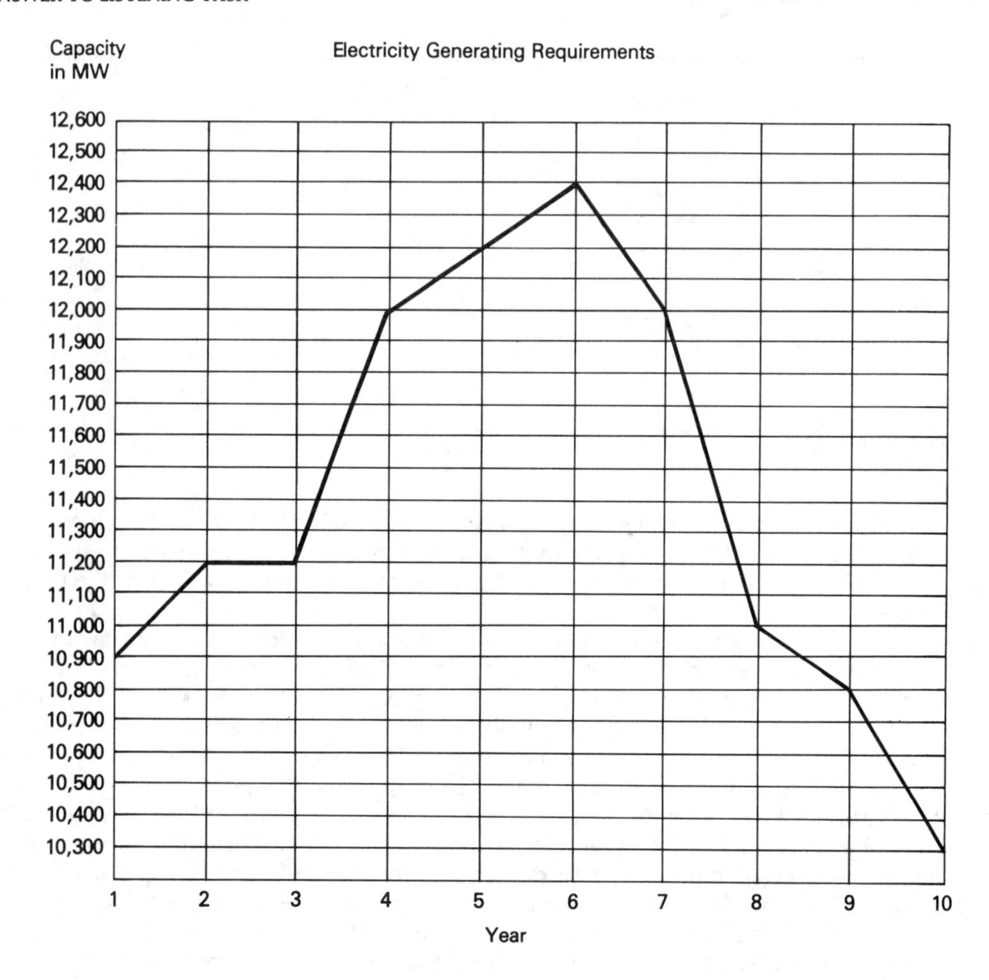

---

3 Controlled Practice

1. In the 1st year capacity *stood at* 16000 MW.
2. In year 2 it *decreased/dropped/fell* by 800 MW.
3. In the 3rd year it *remained* constant *at* 15200 MW.
4. Then in the 4th year it *decreased/dropped/fell to* 14800 MW.
5. And in year 5 we saw a small increase *of* 200 MW.
6. But in the 6th year capacity *decreased/dropped/fell* to 14800 MW.
7. This trend continued, and in the 7th year capacity *decreased/dropped/fell* substantially *by* 800 MW.
8. There was a further *decrease/drop/fall of* 200 MW in the 8th year.
9. And in year 9 capacity *went down to* 13400 MW.
10. But in year 10 capacity showed a small *increase/rise of* 200 MW.

4 Transfer

PAIRWORK

Student B: Your partner is going to give you some information about world electricity production. As you listen to the information, complete the graph below. After you have completed your graph, compare it with your partner's.

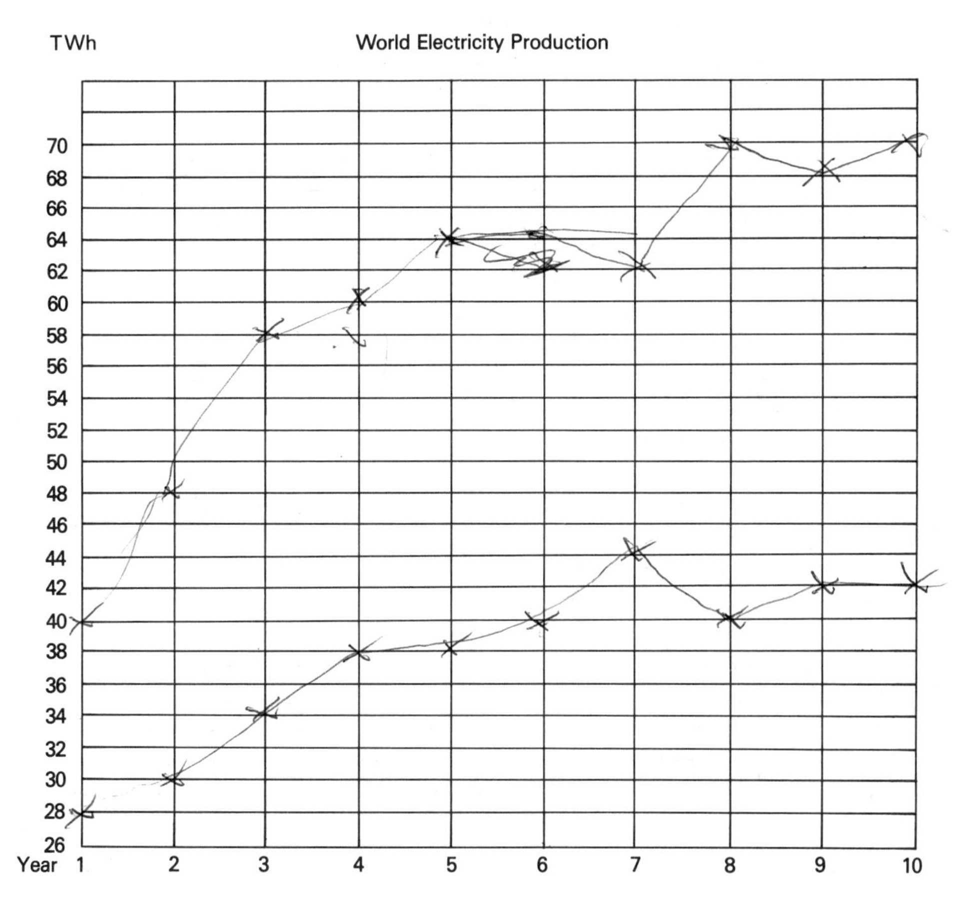

TWh = Terra Watt hours

Student B: Now look at the information below about world electricity production. Then describe the trends to your partner, who will complete a graph according to the information you give.

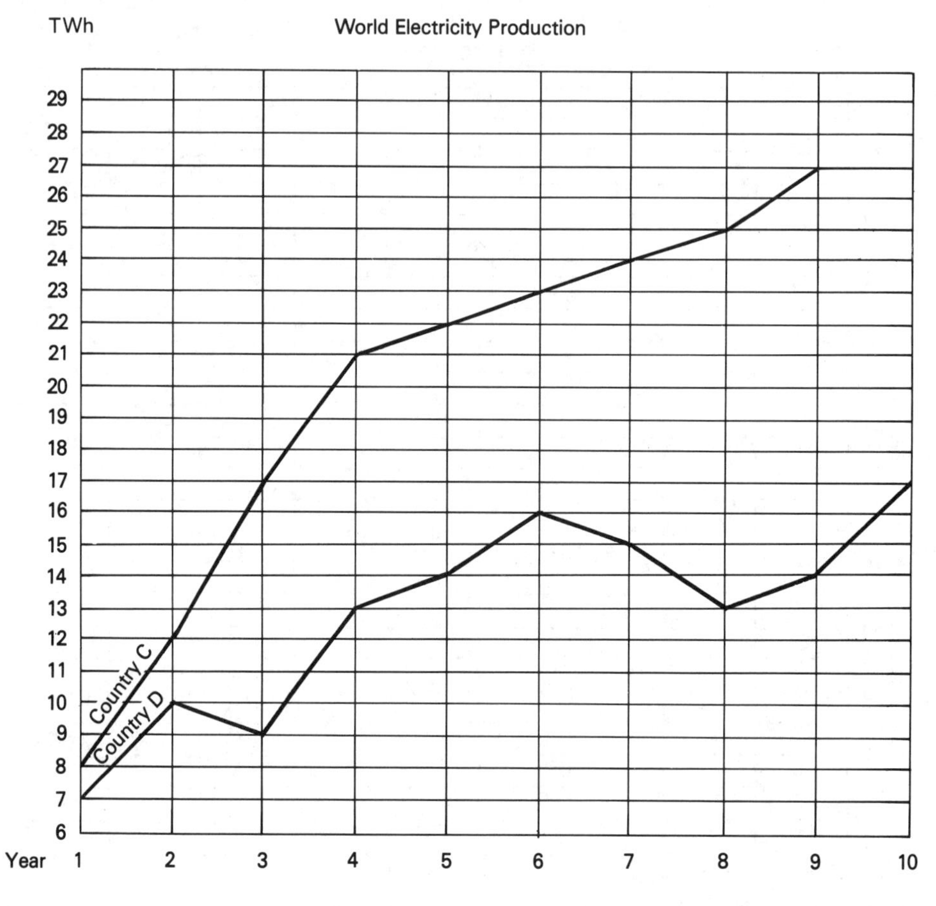

Twh = Terra Watt hours

UNIT 13 Microchip Manufacture

1 Listening

TAPESCRIPT

Before going round the factory, let's have a look at this flow diagram. It should give you an idea of the main stages of the manufacturing process. You can see that we start with silicon rods . . . these are from 4 to 6 inches in diameter. First these rods are cut into thin slices . . . we call these slices wafers . . . and then the faces of the wafers are polished. Next the faces are covered with something we call photoresist — this is a sort of plastic which is sensitive to light. So the faces or wafers are covered with photoresist before entering the photographic part of the process. Then, at the next stage the wafer is exposed to the image from a mask plate — the plate is really a printed diagram of the circuit and you can see the set-up with a light, some lenses and then the mask.

Once the image is on the wafer, it is developed photographically — this means the exposed photoresist hardens and the unexposed photoresist is removed. Now, we come on to the next part of the photographic process . . . having removed the unexposed photoresist, chemicals are applied to process the wafer through the photoresist image. Finally, the photoresist is removed and then it starts all over again . . . the process is repeated many times for other images before sending the wafers for testing and mounting. Anyway, let's go and look in the factory . . .

ANSWERS TO LISTENING TASK

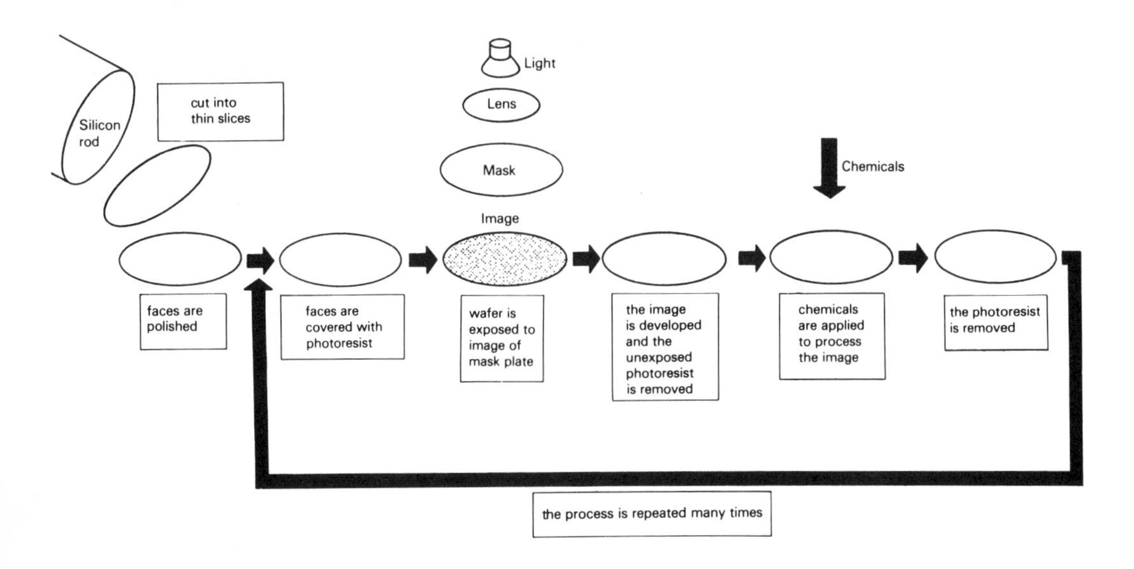

3 Controlled Practice

A. 1. The rods are cut into thin slices.
2. The faces of the wafer are polished.
3. The faces are covered with photoresist.
4. The wafer is exposed to the image of a mask plate.
5. The image is developed.
6. The unexposed photoresist is removed.
7. Chemicals are applied to process the wafer.
8. The photoresist is removed.

B. 1. First/To start with
2. Then/Next
3. After
4. Once
5. Now
6. Once
7. Finally
8. Now

4 Transfer

Student B: Use the flow chart below — the second part of the manufacturing process of microchips — to answer your partner's questions.

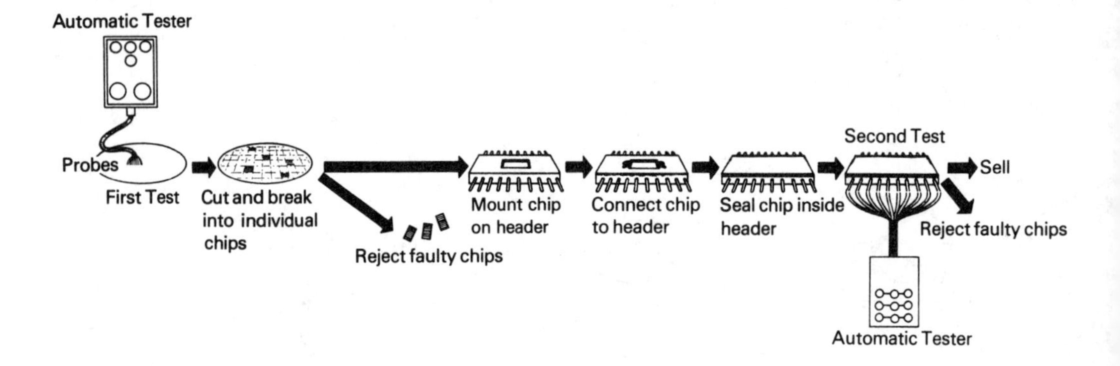

UNIT 14 VCR Recording

1 Listening

TAPESCRIPT

SWITCHBOARD: Video and Computer Centre. Good afternoon.
CUSTOMER: Good afternoon. Could I speak to one of your technical staff, please?
SWITCHBOARD: For video or computers, sir?
CUSTOMER: Video, please.
SWITCHBOARD: One moment, please.
TECHNICAL ASSISTANT: Hello. Can I help?
CUSTOMER: Yes . . . I bought a Tanuyi video cassette recorder from you earlier today, but when I got home I found that the instructions were missing. Well, I've managed to connect the machine to my TV, but I've been trying for one hour to set it up for unattended recording. But I just can't figure it out.
TECHNICAL ASSISTANT: OK. What did you do first?
CUSTOMER: First I pressed STANDBY.
TECHNICAL ASSISTANT: I see. Well, first of all, you shouldn't have pressed STANDBY; you should have pressed TIMER. If you had pressed TIMER, you would have seen the message BLOCK 1 on the VCR display.
CUSTOMER: I see. So first I should have pressed TIMER.
TECHNICAL ASSISTANT: Yes, that's right. And then you ought to have pressed it again.
CUSTOMER: You mean I oughtn't to have pressed STANDBY at all?
TECHNICAL ASSISTANT: That's right, sir. As I was saying, you should have pressed TIMER again. If you had pressed it a second time, you would have seen START on the VCR display. And then you could have selected the starting time of the programme you wanted to record.
CUSTOMER: Yes, I see.
TECHNICAL ASSISTANT: Then you ought to have pressed TIMER a third time . . . and you would have seen STOP on the VCR display. Then you could have selected the finishing time of the programme.
CUSTOMER: Uh-huh.
TECHNICAL ASSISTANT: And if you had pressed TIMER a fourth time, you would have seen PROG (meaning Program) on the VCR display. Then you could have selected the TV channel of the programs you wanted to record.
CUSTOMER: So I should have pressed TIMER four times?
TECHNICAL ASSISTANT: No, sir. You should have pressed it five times. If you had pressed it one more time, you would have seen the actual time and BLOCK 1 on the VCR display . . . and it would have been set up. It's all very simple really.
CUSTOMER: OK, I see. But it would have been much easier if I had had the instruction manual.

ANSWERS TO LISTENING TASK

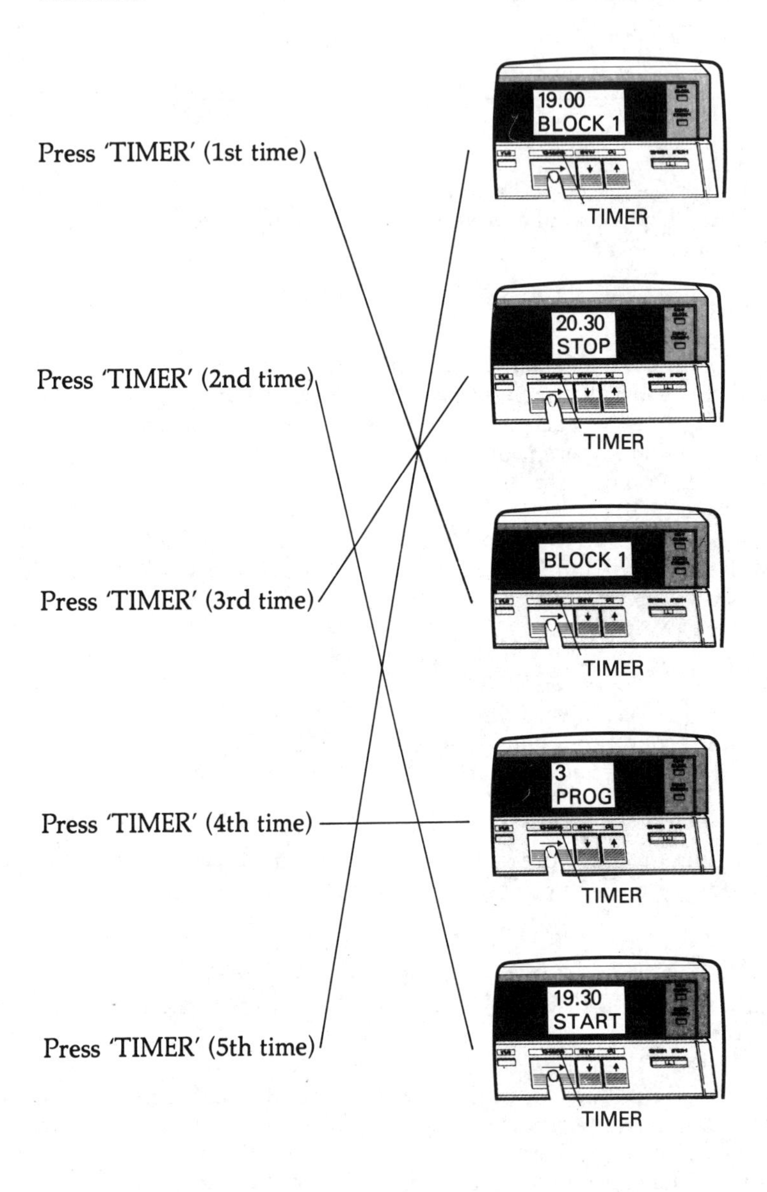

3 Controlled Practice

1a. I should/ought to have read the operating manual.
 b. If I had read the operating manual, I would have seen the section on voltage regulation.

2a. I should/ought to have read the section on voltage regulation.
 b. If I had read the section on voltage regulation, I would not have set it to 110 volts.

3a. I shouldn't/oughtn't to have left it on 110 volts.
 b. If I hadn't left it on 110 volts, it would have worked.

4a. I shouldn't/oughtn't to have opened the case.
 b. If I hadn't opened the case, I wouldn't have got an electric shock.

5a. I should/ought to have switched off the power first.
 b. If I had switched off the power first, I would still have had my VCR.

6a. I shouldn't/oughtn't to have tried to be a technician.
 b. If I hadn't tried to be a technician, I wouldn't have lost a lot of money.

4 Transfer

PAIRWORK

Student B: You are the project co-ordinator responsible for the implementation of a plan to manufacture and transport chemicals. Although your project manager gave you a critical path analysis (activity flowchart) to follow, you modified it because you thought you could save time. Below is the plan you designed for the activities involved. The lowest number (1) shows the activity you did first. (2) is the activity you did next, etc. Numbers which are the same show activities you did at the same time.

Specify Truck Chassis 1	Requisition Chassis 2	Order Chassis 3	Supplier Manufactures Chassis 5	Supplier Delivers Chassis 7		
Specify Instrument Design 1	Requisition Instruments 2	Order Instruments 3	Supplier Manufactures Instruments 6	Supplier Delivers Instruments 8		
Order Sample Chemicals 2	Check Quality of Chemicals 4	Approve Chemicals 5	Place First Order for Chemicals 7	Supplier Delivers Chemicals 9	Store Chemicals 10	
			Start Plant Tests 9	Review Tests 10	Approve Start-Up 11	Start Up Plant 12

Unfortunately your project manager (Student A) has noticed that you have modified the plan. He calls you to his office, and asks you to tell him the order in which you implemented the steps. Use the information above to answer Student A's questions.

You start as follows:

A. So what did you do first?
B: First I specified the truck chassis, and at the same time I specified the instrument design.

UNIT 15 Data Communications

1 Listening

TAPESCRIPT

ADMINISTRATIVE MANAGER: So you think we should change the system?

COMMUNICATIONS ENGINEER: Yes, I do. We're having a lot of problems with the existing configuration.

ADMINISTRATIVE MANAGER: So, what do you suggest?

COMMUNICATIONS ENGINEER: Well, at the moment, we're using a modem linked up direct to our microcomputer and the telephone line. Very simple, in theory we send files from our computer direct to our clients' computers using a software package called 'Communicate'.

ADMINISTRATIVE MANAGER: OK. Well, sounds fine. What's the problem?

COMMUNICATIONS ENGINEER: Well, the weak link is the line. As you know, a lot of our clients are overseas and we're having problems with the lines. The files are being transmitted, but they're not arriving in the same form. The data is being corrupted.

ADMINISTRATIVE MANAGER: What does that mean?

COMMUNICATIONS ENGINEER: Well, say we send a 10 page contract to a client in France. We make the connection, start transmitting and then there's a fault on the line and they receive the contract with some parts missing, or parts they can't understand.

ADMINISTRATIVE MANAGER: That sounds bad.

COMMUNICATIONS ENGINEER: Yes, it is. Even more serious is that it's costing us a lot of money in call charges. Call charges on international lines are high and each time we send a report or contract — let's say 10 pages long — it takes up to 5 minutes to transmit — longer if we have problems with the line.

ADMINISTRATIVE MANAGER: Yes, that is serious. So what do you suggest?

COMMUNICATIONS ENGINEER: Well, I think we should use an electronic mailbox.

ADMINISTRATIVE MANAGER: How does that work?

COMMUNICATIONS ENGINEER: Very simply really. We subscribe to a service called DIALCOM. We send our files to a central computer. The files are stored there and our clients can get the file out when they want. We use our existing equipment and so we only have to pay for the subscription and the call charges.

ADMINISTRATIVE MANAGER: How expensive is it?

COMMUNICATIONS ENGINEER: It's cheaper, but more important — the system — DIALCOM — uses a data network not the normal telephone lines; so transmission is faster and more reliable — there's less chance of the data being corrupted.

ADMINISTRATIVE MANAGER: Right, I think we should go ahead.

ANSWERS TO LISTENING TASK

a. our micro
b. modem
c. telephone line/network
d. modem
e. client's micro
f. data network
g. central computer
h. data network

3 Controlled Practice

A. 1. . . . send . . . costs . . . .
2. . . . is costing . . . .
3. . . . isn't working . . . .
4. . . . accesses . . . wants.
5. . . .'m having . . . .
6. . . . transmit . . . gets . . . .
7. . . .'re spending . . . .
8. . . . don't happen . . . .
9. . . .'re losing . . . .

B. 1. . . . operates . . . .
2. NO CORRECTION NECESSARY
3. . . . control . . . gives . . . .
4. . . . provides . . . .
5. . . . is rapidly expanding . . . .

4 Transfer

Student B: You have recently started using a modem for data communication. You are having problems with it. Get in touch with the modem manufacturer (Student A). He/She will ask you questions about how you are using it. Use the information below to answer these questions:

E.g. Student A: What operating speed are you using?
Student B: We're using 300 bits/sec.

MODEM TYPE: V22
OPERATING SPEED: 300 bits/sec
MODE OF OPERATION: Call mode
Half duplex
CONNECTION TO COMPUTER: Parallel cable — RP 354
CONNECTION TO LINE: Telephone socket
OPERATING CONNECTION: Manual dialling

This information tells you how you *are using* the modem at the moment. Student A will tell you how you should use it.

UNIT 16 Information Retrieval

1 Listening

TAPESCRIPT

A: Well, if I can have your attention for a moment. At the moment we've got all our records on microfilm. When one of our clerks wants to look at a document, he goes down into the microfilm library, inserts the microfilm index into the microfilm reader, selects the relevant microfilm, searches in the reader for the document and then, if he wants, he can take a thermal copy of the document. I'm sure you'll all agree that the process is slow and inefficient.

So what we want to do is update the system. First of all, we're going to install work stations in all the main offices.

B: What will a work station look like?

A: You can see on this diagram here, a work station will consist of a terminal with its own keyboard and then a connection to a high-quality printer. The operator will call up the index on his terminal. The index will now be on disk in a central computer. This central computer will be connected to a central microfilm reader. The reader will be down in the microfilm library. Now the new thing here will be a microfilm autoloader. This will select the right microfilm from the library and then insert the film into the reader. The image will be electronically scanned and then transmitted via the central computer to the work station. The operator will view the document on his terminal screen and then, if he wants, take a top-quality hard copy from the printer.

B: What's the time schedule on this?

A: Well, as I said, we're going to install the work stations in the first phase. We plan to do this by the end of the year.

B: What about the autoloader?

A: Well, at the moment we're doing a trial. If all goes well, we plan to install it at the beginning of next year. The central computer is already on site. We're working on some new software which should be ready soon.

B: So what about the whole system?

A: We expect the whole system to come on line in spring next year.

B: I see. Can you tell us some more about . . .

ANSWERS TO LISTENING TASK

a. microfilm library
b. microfilm autoloader
c. central microfilm reader
d. central computer

e. keyboard, f. terminal, g. printer } workstation

3 Controlled Practice

1. ... goes ... .
2. ... inserts ... .
3. ... selects ... .
4. ... searches ... .
5. ... wants, ...

6. ... will call ... .
7. ... will be ... .
8. ... will select ... .
9. ... will be ... scanned ... .

10. ... will view ... .
11. ... are doing ... .
12. ... are working ... .

13. ... are going to install ... .
14. ... plan to install ... .
15. ... expect ... to come on line ... .

4 Transfer

Student B: Listen to Student A's presentation about the 'present site' of information flow in your company. He/She will also tell you about current work. Then present to him/her the 'future state' (see diagram below) and the project timing.

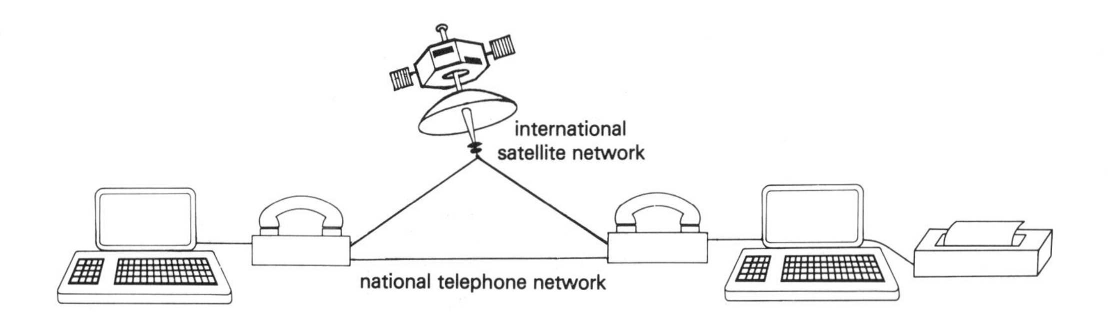

Project timing:
Install work stations in head office: January next year
Install work stations in sales branches: April next year
National test for electronic mail: June next year
International test for electronic mail: July next year
System on line (national and international network): September next year

UNIT 17 International Aviation Standards

1 Listening

TAPESCRIPT

Recent years have seen many developments and improvements in the fields of aircraft design, instrument design and flying procedures. But these developments and innovations must not lose sight of 6 key objectives – 6 goals which are essential for the future of our industry. And this afternoon I would like to present these 6 points and explain why, in my opinion, they are so important for us in the area of aviation.

Our first goal is efficiency. We are a service industry and, like all service industries, we must operate efficiently in order to survive. We must be efficient in all aspects of the services we provide, but especially in our Air Traffic Control Service. This service is central to the smooth and safe operation of our activities. And I believe that we must at all times try to improve our air traffic control efficiency.

Secondly we have safety. If our Air Traffic Control Service operates efficiently, we can expect to fly safely – and it is safe flights that our clients expect. The flights may be delayed ... passengers will complain and grumble, but they will accept it. Accidents, however, affect our image very badly. And passengers are well aware of safety records. Now, we do have a good flight safety record at the moment, but that doesn't mean it can't be improved.

Now on to the next point – speed. Advances in aircraft engineering have shrunk the world. I mean that supersonic travel enables passengers to fly quickly between countries and continents. In our modern business world the senior executive needs fast flights, and therefore speed is an important aspect of our service. Speed is also an important factor in relation to our Air Traffic Control Service. It needs to respond quickly in many different types of situations. And our new generation of aviation equipment enables us to do just this – to provide information quickly to planes on the ground and in the air. Staying with aviation instrumentation for the moment, we must not forget the role and importance of precision. All flying instrumentation, both on the ground and in the aircraft, must give precise information. By providing information precisely we improve our service in terms of speed, efficiency and safety. But, flight instruments also need to be reliable; in addition, the personnel who operate them must work reliably, too; and the information they provide must achieve a high degree of reliability.

My final point this afternoon is comfort. Passengers want not only a safe, fast and efficient service, they also want to travel comfortably. An uncomfortable flight is always remembered, and the information passed on – to our disadvantage. So, we must aim to provide comfortable flights for our passengers through a caring and courteous approach to their requests and problems.

Well, I hope that we all agree in principle with the objectives, and that we will continue to make improvements in all the areas – to provide a better service to our clients.

ANSWERS TO LISTENING TASK

Goals	*Areas*
efficiency	Air Traffic Control Service
safety	flights/(Air Traffic Control Service)
speed	flights/Air Traffic Control Service/(modern business world)
precision	instrumentation/information
reliability	instruments/personnel/information
comfort	flights

( ) = possible additional answers

3 Controlled Practice

1. uncomfortable
2. dangerous
3. substantially
4. reliable
5. quickly
6. efficiently
7. precisely
8. safety
9. occasionally
10. usually
11. acceptable
12. essential
13. quickly
14. possible
15. normal
16. comfort
17. high
18. efficiency
19. precision
20. quickly

UNIT 18 Scuba Diving

1 Listening

TAPESCRIPT

This is basically a new plan for a computer for divers. It will provide in one instrument all the functions that are currently on a diver's watch — depth meter and air contents gauge — plus other data. It is to be used together with official divers' tables, which give information about safe diving limits. Now, we expect the device will consist of a display section, which will provide visual information and audible warnings. Firstly, there is 'WATER TYPE' information. If a diver wants an accurate depth reading, he must set the appropriate water type. If he is in the sea, he will set 'seawater'; otherwise he will set 'freshwater'. Then there is 'TIME' information. It is the time from the beginning of the diver's descent to the end of his ascent. So, if a diver wants to know how long he has been down, he can see this from the display. After that there is the 'STOP WATCH' information. For example, if a diver needs to spend five minutes at a depth of 5 metres for decompression, he will start his stopwatch, and will wait until the five minutes have elapsed. Then we have information about 'AIR TANK CONTENTS' — measured in bars. A bar is a metric measurement approximately equal to one atmosphere of pressure. So, if a diver wanted to know the amount of air left in his tank, he could see this from the contents display. After that there is 'DEPTH' information measured in metres. And this shows the diver's current depth. Then we will show 'BATTERY LEVEL' information. It will show the proportion of power left in the batteries. It will indicate if the level is low, medium or high. Of course, if the batteries are low, the instrument will not function at all. The rest of the information is transmitted via audible warnings. Now, if he hears a warning signal, the diver should look at the instrument and he will see the relevant area of the display flashing. The '50 and 30 BARS WARNING' display will show him when the air contents of his tank reach those levels. Then we have the 'PRE-ASCENT WARNING', which tells him when he must start his ascent. Before he descends, the diver sets the time when he must start his ascent. He does this by referring to official diving tables. If he began his ascent before this display started flashing, he would be within safe limits; if he didn't ascend then, he would expose himself to decompression sickness. If he then started his ascent and went up too quickly, he would see a warning light — 'ASCENT TOO FAST'. So this instrument will provide both information about time, depth, and air contents, and will warn about critical conditions.

Action/Situation	Visual information	Audible warning
1. Go to the surface now.		√
2. You are 50 metres below the surface.	√	
3. Battery level — OK.	√	
4. Seawater dive.	√	
5. You started your descent 10 minutes ago.	√	
6. Look at your AIR TANK CONTENTS display.		√
7. Slow down your ascent.		√
8. The amount of air in your tank is 144 bars.	√	
9. You started to decompress 4.28 minutes ago.	√	

3 Controlled Practice

1. If you are not under water, the air pressure will vary according to your altitude.
2. If you were under water, the pressure would increase.
3. If the pressure increased, you would use your air more quickly.
4. If you are under water, you will need to know exactly the air time left.
5. If you stayed under water for a long time, you would need to spend time in decompression.
6. If you don't spend time in decompression, you may suffer from the bends.
7. If you want to avoid decompression, you must start your ascent within one minute.
8. If you didn't start your ascent then, you would have to stay at 5 metres for 10 minutes.
9. If you look at your diver's watch, you will see your present depth in metres.
10. If you saw the 'ASCENT TOO FAST' display, what would you do?

UNIT 19 Control Systems

1 Listening

TAPESCRIPT

TECHNICAL SALES REP: Well, your first decision is whether to go for cab control or floor control. Let's go quickly through the cab options. Basically cabs make sense where the cranes are travelling at speed, or where the operating conditions make it necessary to protect the operator.

CUSTOMER: What sort of operating conditions are you talking about?

TECHNICAL SALES REP: Well, extreme heat or extreme cold, dangerous or unpleasant fumes — that sort of thing.

CUSTOMER: I see and the cab is standard?

TECHNICAL SALES REP: No, there are 4 basic positions. There's the laterally mounted cab — here you can see it at one end of the crane bridge. This position gives good visibility of the load. Then, if you look here, there's the centrally mounted cab — here it is in the centre of the crane bridge. Then there's the crab mounted cab — you can see it here mounted directly over the load — this position is very good for close supervision of the load and finally an independently mobile cab which moves along the crane bridge and so the operator can vary his distance from the load.

CUSTOMER: I see. And you mentioned a cheaper alternative?

TECHNICAL SALES REP: Yes, that's floor control. Usually this means using a control pendent. Here you can see a mobile control pendent. This means the operator can move close to the load if he wants to guide it manually, or he can operate it at a safe distance. In this picture here, you can see a crab suspended control pendent. In this case, the operator is always close to the load and can guide it manually. We recommend this type for repair and assembly work where positioning is vital. The other extreme is the laterally suspended control pendent — this can be used where direct manual guiding isn't necessary.

CUSTOMER: Is that it?

TECHNICAL SALES REP: Not quite. Finally there's the remote control option. Either using a stationary control console or a small remote control device. These are extremely useful where the operator can't work in the crane area — for example when the crane is handling dangerous chemicals or when the crane must be controlled from several different points.

CUSTOMER: I see.

TECHNICAL SALES REP: Anyway, let's go and look at some of the cranes in action.

ANSWERS TO LISTENING TASK

1. d
2. g
3. f
4. a
5. b
6. e
7. h
8. c

3 Controlled Practice

A. 1. extremely hot operating conditions
   2. a centrally mounted cab
   3. a manually guided control pendent
   4. a stationary control console
   5. substantially improved visibility

B. 1. Extremely unpleasant operating conditions
   2. A heavy load industrial application
   3. A well qualified electronics engineer
   4. A robotically controlled chemical process
   5. A cheaply manufactured device
   6. (Specially developed) customised software (NOTE: 'customised' = 'specially developed')

4 Transfer

Student B: Answer Student A's questions using the information below. Together try to build up a noun phrase to describe the object.

e.g. A car
Student A: What type? You: Sports
Student A: What size? You: 2000 cc
Student A: Where is it made? You: Japan

⟹ *A 2000 cc Japanese sports car*

1. *A machine*
type: welding
size: large
Operation: automatic

⟹ _

2. *A drill*
type: industrial
Operation: driven by electricity

⟹ _

3. *A computer*
type: micro
size: 32 bit
Origin: America

⟹ _

NOTE: If you want, you can add more information!

UNIT 20 Project Planning

1 Listening

TAPESCRIPT

JIM: Hi, Chris. It's Jim here. I'm calling about the Saudi project ... to find out how the work's coming along.
CHRIS: Not bad, we're mostly on schedule.
JIM: Has all the equipment been installed?
CHRIS: Yes, we finished installation last week. We start testing the machines on Monday next week.
JIM: How long will that take?
CHRIS: Well, we've scheduled three weeks so we should finish at the end of the month.
JIM: Good. What else?
CHRIS: Well, the operator training has already started. We kicked off on Wednesday this week and the first course ends next Friday.
JIM: Oh yes, that was one of the things I wanted to mention. Fred Hyman, the maintenance trainer should arrive at the weekend.
CHRIS: Fine, do you know what time?
JIM: No, but I expect he'll arrive at 12 on Saturday. I'll telex you as soon as I know for certain.
CHRIS: OK. Anyway he'll have a week before he starts training. The first maintenance course is due to begin a week from Monday.
JIM: When do you plan to finish the training programme?
CHRIS: Just a moment, I'll look at the planner ... here it is, um, ..., the last course is in July – that's the Supervisors' course – if all goes well that'll finish at the end of the month and they'll be ready to start work at the beginning of August.
JIM: So you plan to start up in August?
CHRIS: Yes, if all the tests are OK, we've got a provisional start-up date on 25th August ... for the first two weeks we'll be building up capacity slowly ... hope to reach full capacity by September 8th.
JIM: Right, that's the other thing I wanted to mention. The client wants an official opening date for the plant – when do you suggest?
CHRIS: Well I've been talking to some of the Saudis here – in fact I talked to the Works Manager a couple of days ago – he reckoned the middle of September would be fine. Let me just look at my diary ... shall we say September 15th?
JIM: Sounds fine. Anything you need?
CHRIS: Um. I don't think so. Oh yes. Could you send some more copies of the Operators' Manual. Let's say about 20.
JIM: Of course. I'll send them off today. If I get them off airmail they should be with you by Monday.
CHRIS: Right thanks Jim.
JIM: You're welcome. Speak to you again soon.
CHRIS: Yes, Bye.
JIM: Bye.

ANSWERS TO LISTENING TASK

a. installation of equipment
b. testing of machines
c. operator training
d. maintenance course
e. supervisors' course
f. provisional start-up date
g. full capacity date
h. official opening date for the plant

3 Controlled Practice

A. 1. last
2. on . . . next
3. in 3 weeks
4. on . . . this
5. next/at the end of next
6. from Monday
7. at the end of
8. on
9. by
10. on

B. I confirm F. Hyman arrives at Riyadh on Saturday at 1200. He expects to start training on 17th June. Please meet him at the airport. 20 operators' manuals were sent yesterday. They should arrive at the beginning of next week.
Regards
Jim Coleman

4 Transfer

Student B: Use the planner below to answer Student A's questions.

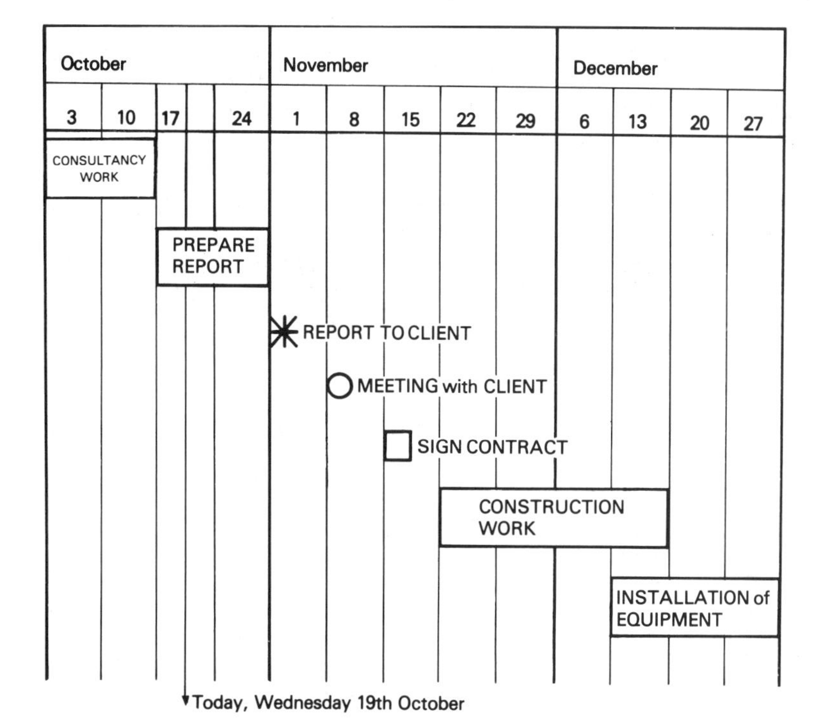

UNIT 21 Speeding up Air Traffic

1 Listening

TAPESCRIPT

The new computer system that we have installed handles 3 areas of need for passengers — flight information, passenger check-in and duty-free sales. First flight information. Over 50 visual display boards have been installed inside the airport — in the main concourse, in the arrival and departure lounges, at arrival and departure gates, and in the baggage collection areas. This information is also repeated on the airport's closed-circuit TV system. The characters on the boards are generated electronically, and therefore we can generate both standard flight information and non-standard messages. After the departure of a flight the details are automatically taken off the boards. Each message is composed of a matrix of small holes. Over each hole there is a metallic disc. A character is thus formed from back-projected light shining through the required holes, after the discs have been removed. This system is much more flexible and reliable than the alternative flap system.

Secondly, passenger check-in. Our new system represents a move away from individual airline check-in to integrated check-in. We have 14 check-in points along the whole of one side of the main concourse. And a passenger can go to any check-in point.

Finally, our duty-free shop is in the departure lounge. When a passenger walks into the shop, he can select from a wide range of goods. Each item on sale includes the price and stock number, coded onto the price tag. At the point-of-sale terminal the data is read by moving the price identification bar code across a light beam, and the terminal produces a 3-part receipt. One copy goes to the customer; one to the shop; and one to Customs. The shop's data is stored on disk, and then sent to head office for processing.

So a passenger can see his flight details on our new visual display boards, check-in at any of our new check-in points, and go to the duty-free shop, where he can select from a wide range of goods. Then he can go out of the shop and into the lounge until he sees a departure message on the visual display board. And it's all possible as a result of our new computer system.

Facilities (handled by the computer system)	*Where*	
50 Visual Display Boards have been installed	in	the main concourse
	in	the arrival and departure lounges
	at	arrival and departure gates
	in	the baggage collection areas
Flight information is also displayed	on	the airport's closed-circuit TV system
A character is formed by light shining	through	the required holes
We have 14 check-in points	along	the whole of one side of the main concourse
A passenger can go	to	any check-in point
Our duty-free shop is	in	the departure lounge
Each duty-free item has the price coded	onto	the price tag
One copy of the receipt goes	to	the customer
After buying his duty-free goods a passenger can go	out of	the shop and
	into	the lounge

3 Controlled Practice

A.
1. When you arrive inside (L) the airport building, go to (M) the check-in desk.
2. Make sure you have taken your ticket, passport and money out of (M) your baggage, before you put it on (M) the scales.
3. The clerk at (L) the desk will check your ticket.
4. She will tie a baggage label onto (M) the handles of your suitcases.
5. Then you can either wait in (L) the main concourse or go through (M) passport control to (M) the departure lounge.
6. At (L) most airports you will find a duty-free shop and a restaurant in (L) the departure lounge.
7. Do not take the seal off (M) your duty-free purchases before you get on (M) the plane.
8. Once inside (L) the plane, stow your hand baggage safely: do not put any heavy bottles into (M) the overhead lockers.

B.
1. This is Flight 307 *to* Madrid.
2. Hold *at* Gate Two.
3. Taxi *into/to* position and hold.
4. Wait for further instructions *from* the tower.
5. Proceed to holding point. Clear for take-off. Start taxiing *along* the runway.
6. Contact departure control *on* frequency 128.0.
7. Climb *to* flight level 120.
8. You are now clear to move *outside/out* of British controlled airspace into French airspace.
9. You are about to pass *through* moderate turbulence.
10. We are *out of* turbulence now. Clear conditions ahead.

UNIT 22 Electronic Assembly

1 Listening

TAPESCRIPT

A: We've heard a lot about surface mounting recently. What exactly is it?

B: Well, normally when you want to assemble a circuit board for a television or a computer, holes are drilled in the board and the components — resistors, capacitors, chips and so on — or rather the wire leads from the components — are pushed through these holes and soldered to copper pads on the other side. Now, surface mounting uses components already manufactured with pads or tabs on the underside or edge and these are soldered directly to the circuit pads.

A: I see. So what are the advantages?

B: Well firstly size. The old components were much bigger and heavier. They also had to be more robust in order to tolerate the autoinsertion through the holes. Surface mounting is a much gentler process and the spacing between the components can be as little as 0.05 inch. All in all, the board, in some cases, can be four times smaller and much lighter too.

A: But surely these new components must be more expensive?

B: That's true. At the moment they are about 1½ times more expensive but that's only because there's no high volume production yet. But, in fact the actual assembly process is both faster and cheaper.

A: So, all pcb manufacturers and assemblers are likely to change to surface mounting?

B: I think so. Especially since they're more reliable than conventionally mounted components — there are no more wire leads which can vibrate and lead to faults. The placement of the components is also ten times more accurate.

A: Are there any disadvantages besides the cost of the components?

B: Well, at the moment, they're more difficult to autotest but I'm sure that's only a temporary problem.

ANSWERS TO LISTENING TASK

Features	*Surface-mounted versus conventional assembly*
Size	√
Weight	√
Cost of production	×
Speed of assembly	√
Cost of assembly	√
Reliability	√
Accuracy of placement	√
Autotesting	×

3 Controlled Practice

A.

Feature	Adjective	Opposite adjective
Weight	heavy	light
Size	big	small
Width	wide	narrow
Ease	easy	difficult
Speed	fast/quick	slow
Reliability	reliable	unreliable
Accuracy	accurate	inaccurate
Efficiency	efficient	inefficient
Cost	cheap	expensive
Strength	strong/robust	weak/gentle

B.
1. bigger
2. lighter
3. cheaper
4. faster/quicker
5. more expensive
6. more reliable
7. more accurate
8. easier
9. narrower
10. stronger/more robust

4 Transfer

Student B:

	SM board	Conventional board
Size		1: 22 in. w: 10 in
Thickness		1.5 cm
Weight		780 gms
MTBF (mean time between failure)		850 hours
Accuracy of component placement:		± 0.01 in
Testing ratio (autotest: manual test)		95 : 5
Assembly method autoinsertion: machine insertion: manual insertion:		 30% 60% 10%

UNIT 23 Energy Sources

1 Listening

TAPESCRIPT

A: Well, what do you want to know about energy generation?
B: We hear so much these days about different fuels and processes. We are told that nuclear power is more efficient than conventional fossil fuels. And we know that fossil fuels are limited. How can we compare the efficiency of the different fuels and processes?
A: Well, first of all, what types of fuel do you know?
B: Conventional fossil fuels – that is, oil, gas and coal – and nuclear fuels – that is, uranium and plutonium.
A: Right, and what processes do we use?
B: Well, I know that there are different nuclear reactors and different conventional processes.
A: Well, let's imagine a bucket of fuel.
B: What exactly do you mean? How much does a bucket hold?
A: Let's say a bucket holds 10 kilograms.
B: So how long does a bucket last?
A: Well now, that depends on the type of fuel and the type of process. And let's look at a 2 million kilowatt power station.
B: How many megawatts does that make?
A: 2 million kilowatts make 2000 megawatts. OK?
B: OK. So which fuel produces the most energy?
A: Well, that's nuclear fuel.
B: And which process does it use?
A: It uses the most efficient nuclear process, which converts all the matter in this fuel into energy.
B: So how long will it last?
A: Well, you may be surprised when I tell you that it will last eight and a half years. In fact you will be very surprised if you compare it with a hydrogen fusion reactor.
B: How long does a bucket of fuel last using that process?
A: Only 2 weeks.
B: Only 2 weeks. That's certainly an incredible difference.
A: And there's more to come.
B: What do you mean?
A: Well, the next process is a fast reactor.
B: Yes. When will that need more fuel?
A: After just a week. And now we come on to natural uranium.
B: And when will that fuel stop producing energy?
A: After 3 days. Now let's look at conventional fossil fuels, shall we? How long do you think a bucket of oil will last?
B: One hour?
A: Well . . . nearly. In fact it will last one eighteenth of a second! And the same goes for coal.
B: So which country today produces most electricity using nuclear energy?
A: Well, in Europe, France is top and then West Germany.

ANSWERS TO LISTENING TASK

Energy generation per 10 kgs in a 2 million kilowatt power station	
Fuel/Process	*Running time per 10 kgs*
Nuclear Power	8.5 years
Hydrogen Fusion Reactor	2 weeks
Fast Reactor	1 week
Natural Uranium	3 days
Oil	one eighteenth of a second
Coal	one eighteenth of a second

3 Controlled Practice

1. How much energy do we use per day?
2. How do we generate this energy?
3. Why is nuclear power becoming more important?
4. When did we start to use nuclear power?
5. What do we do with nuclear waste?
6. How long have we been dumping nuclear waste in the sea?
7. Where else can we dump nuclear waste?
8. Which country produces most nuclear energy?
9. How many nuclear power stations does that country have?
10. Where does that country dump its nuclear waste?
11. When will fossil fuels run out?
12. How long can we survive without nuclear energy?

UNIT 24 Factory Automation

1 Listening

TAPESCRIPT

SIMON: So, let's summarise our developments over the last 2 years. 2 years ago, our production line was largely manual. We transported basic components to the line by fork-lift truck. About 25 workers manned the assembly line and at the end of the line, we did the packing and sorting by hand. I think you'll all agree, things have changed a lot during the last two years.

Our first step on the road to automation was last year. You'll remember we installed automatic packing equipment and reduced the numbers of workers in the packing department from 6 to 2. As part of the packing line, we also introduced automatic sorting using a bar code reader.

The next step — during this year — we have gradually automated the assembly line itself. We've reduced the number of workers from 25 to 15 and we've invested nearly $½ million in automatic assembly equipment. Right, very briefly that just about brings us up to date. I'd like you to listen to John now. He's going to talk about the third and final phase.

JOHN: Right, thanks Simon. As Simon has said, we've successfully automated the second two parts of the production line — last year the packing and sorting stage, and then this year the assembly line itself. My team has been studying the third phase of automation which is, in fact, the first stage in the line — the supply of raw materials to the assembly line.

As you know, we receive the basic motor from our main supplier by truck. We unload manually and then store the motors here in the stock room, before placing them on the conveyor at the start of the assembly line. The components — we also store here and then take them in trays to the work stations by fork-lift truck. Altogether 8 employees work in this area.

So, we have looked at two solutions. The first is total automation involving automatic picking in the store using bar codes again and then an automatic feeder onto the conveyor. And for the supply of the components in this solution, we will use microtrucks — a sort of automatic guided vehicle — to transport the components to the work stations. If we decide on this solution, we will reduce the workforce in the supply area from 8 to 2.

The second solution is partial automation. Continue with manual picking in the store, but then automate the supply of motors to the assembly line — in other words install an automatic feeder. In this solution we don't plan to automate the supply of components to the work stations. In this case, we would reduce the workforce from 8 to 6. So, it's a difficult . . .

ANSWERS TO LISTENING TASK

1. A
2. A
3. A
4. B
5. B
6. B
7. C
8. C
9. C
10. C
11. D
12. D

3 Controlled Practice

A. 1. manned/worked on
2. reduced
3. 've (have) reduced
4. work
5. 'll (will) reduce

B. 1. did
2. installed
3. introduced
4. 've (have) automated
5. 've (have) invested
6. unload
7. 'll (will) use

4 Transfer

Student B: Student A will tell you about the product development of products A and B. He should answer these questions:

What did the product look like?
How did it work?
How was it made?
How has it changed?
How has it improved?
What haven't they developed (yet)?
How is it made?
How will it change in the future?

Then you should tell him/her about products C and D.

Product C

Product D

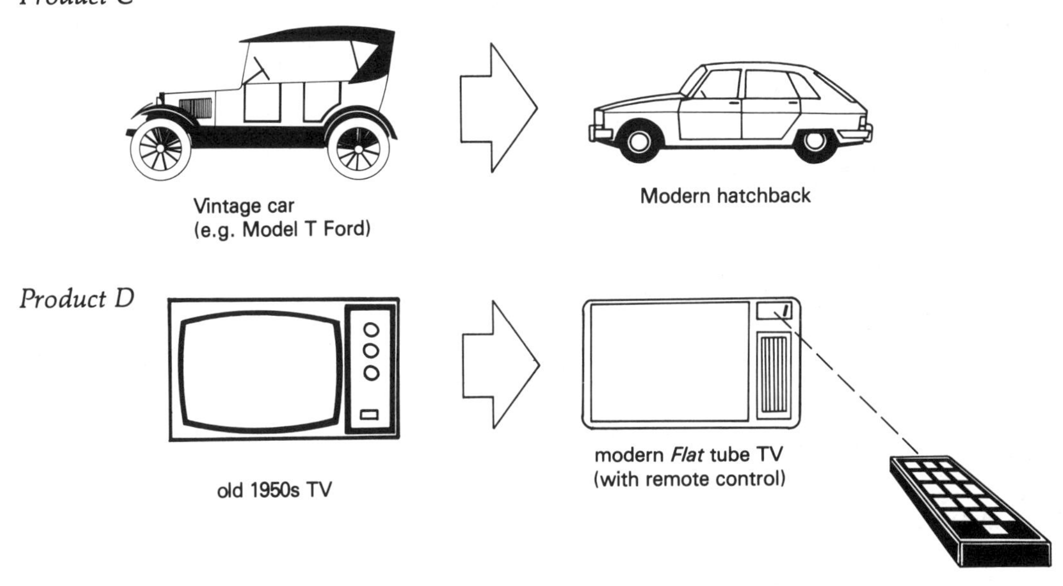

VOCABULARY INDEX

n = noun
v = verb
adj = adjective